CalcLabs with Maple
Single Variable Calculus

FOURTH EDITION

Philip B. Yasskin
Maurice Rahe
David Barrow
Art Belmonte
Albert Boggess
Jeffrey Morgan
Kirby Smith
Michael Stecher

Texas A&M University 2009

BROOKS/COLE
CENGAGE Learning

Australia • Brazil • Japan • Korea • Mexico • Singapore • Spain • United Kingdom • United States

BROOKS/COLE
CENGAGE Learning™

© 2010 Brooks/Cole, Cengage Learning

ALL RIGHTS RESERVED. No part of this work covered by the copyright herein may be reproduced, transmitted, stored, or used in any form or by any means graphic, electronic, or mechanical, including but not limited to photocopying, recording, scanning, digitizing, taping, Web distribution, information networks, or information storage and retrieval systems, except as permitted under Section 107 or 108 of the 1976 United States Copyright Act, without the prior written permission of the publisher.

For product information and technology assistance, contact us at
**Cengage Learning Customer & Sales Support,
1-800-354-9706**

For permission to use material from this text or product, submit all requests online at **www.cengage.com/permissions**
Further permissions questions can be emailed to **permissionrequest@cengage.com**

ISBN-13: 978-0-495-56062-3
ISBN-10: 0-495-56062-6

Brooks/Cole
10 Davis Drive
Belmont, CA 94002-3098
USA

Cengage Learning is a leading provider of customized learning solutions with office locations around the globe, including Singapore, the United Kingdom, Australia, Mexico, Brazil, and Japan. Locate your local office at: **www.cengage.com/international**

Cengage Learning products are represented in Canada by Nelson Education, Ltd.

For your course and learning solutions, visit **www.cengage.com/brookscole**

Purchase any of our products at your local college store or at our preferred online store **www.ichapters.com**

Printed in the United States of America
1 2 3 4 5 6 7 12 11 10 09

Contents

Introduction	vii
Syllabi for Calculus I and II	x
Basic Maple Commands	xi

1 Getting Started — 1
- 1.1 Maple as a Calculator — 1
- 1.2 Assigning Variables — 3
- 1.3 Algebra Commands — 4
- 1.4 Plots — 6
- 1.5 Summary — 9
- 1.6 Exercises — 10

2 Expressions, Functions and Equations — 14
- 2.1 Expressions and Functions: Definition and Use — 15
- 2.2 Expressions and Functions: Plots — 20
- 2.3 Equations in One Variable — 24
- 2.4 Equations in Two or More Variable — 28
- 2.5 Summary — 31
- 2.6 Exercises — 32

3 Data Sets and Parametric Curves — 37
- 3.1 Data Points and Curve Fitting — 37
- 3.2 Parametric Curves — 42
- 3.3 Inverse Functions — 44
- 3.4 Polar Curves — 47
- 3.5 Summary — 49
- 3.6 Exercises — 51

4 Differentiation — 54

- 4.1 The Limit of the Difference Quotient 54
- 4.2 Differentiating Functions 56
- 4.3 Differentiating Expressions 58
- 4.4 Implicit Differentiation 59
- 4.5 Linear Approximation 62
- 4.6 Summary 63
- 4.7 Exercises 65

5 Applications of Differentiation — 73

- 5.1 Related Rates 73
- 5.2 Local Extrema 75
- 5.3 Graphical Analysis 76
- 5.4 Designer Polynomials 78
- 5.5 Absolute Extrema on an Interval 80
- 5.6 The Most Economical Tin Can 81
- 5.7 Summary 83
- 5.8 Exercises 86

6 Integrals — 93

- 6.1 Visualizing Riemann Sums 93
- 6.2 The Computation of Integrals 95
- 6.3 Integration by Substitution (Change of Variables) 98
- 6.4 Integration by Parts 100
- 6.5 Integration using Partial Fractions 101
- 6.6 Approximate Integration 102
- 6.7 Summary 105
- 6.8 Exercises 106

7 Applications of Integration — 109

- 7.1 Area 109
- 7.2 Volume 111
- 7.3 Arc Length and Surface Area 114
- 7.4 Introduction to Fourier Series - Cosine Expansions 116
- 7.5 Summary 120
- 7.6 Exercises 121

8 Differential Equations — 125

- 8.1 Explicit Solutions 125
- 8.2 Direction Fields 127
- 8.3 Numerical Solutions 128
- 8.4 Systems of Differential Equations 130
- 8.5 Summary 133
- 8.6 Exercises 133

9 Sequences and Series — 135

- 9.1 Sequences and Their Limits 135
- 9.2 Series and Their Sums 137
- 9.3 Convergence of Series 138
- 9.4 Error Estimates 140
- 9.5 Taylor Polynomials 143
- 9.6 Summary 145
- 9.7 Exercises 145

10 Programming with Maple — 149

- 10.1 Conditional and Logical structures 149
- 10.2 Looping Structures 152
- 10.3 Procedures 155
- 10.4 Additional Programming Constructs 163
- 10.5 Exercises 163

11 Troubleshooting Tips — 165

- 11.1 Missing or Incorrect Punctuation 166
- 11.2 Unmatched Punctuation and Commands 170
- 11.3 Confusion between Functions and Expressions 173
- 11.4 Unexpected Current Values of Variables 177
- 11.5 Using On-Line Help 181
- 11.6 Failure to Plot 185
- 11.7 Confusing Exact and Approximate Calculations 187
- 11.8 Debugging Procedures 188
- 11.9 Forgetting to Save a Maple Session and Losing It 191
- 11.10 Trying to Get Maple to Do Too Much 192

12 Projects — 195

Projects for Calculus I
- 12.1 A Power Relay Station 196
- 12.2 The Search for the Meteorite 197
- 12.3 Speed Limits 198
- 12.4 Terminal Velocity 200
- 12.5 The Ant and the Blade of Grass 201
- 12.6 Distance Between Two Curves 204
- 12.7 Tangent and Normal Lines 205
- 12.8 Parameterizing Letters 206
- 12.9 Parametric Curves 207
- 12.10 Seeing a Blimp 208
- 12.11 Static and Dynamic Tension 209

Projects for Calculus II
- 12.12 Calculus I Review 210
- 12.13 Calibrating a Dipstick 211
- 12.14 The Area of a Unit p-Ball 212
- 12.15 The Center of the State of Texas 213
- 12.16 Center of Gravity of a Parabolic Plate 214
- 12.17 The Oil Tank 215
- 12.18 The Skimpy Donut 216
- 12.19 Area Between a Curve and Its Tangent Line 217
- 12.20 Curves Generated by Rolling Circles 218
- 12.21 The Wankel Rotary Engine 220
- 12.22 Shakespeare's Shylock 224
- 12.23 The Bouncing Ball 226
- 12.24 Pension Funds 226
- 12.25 The Flight of a Baseball 227
- 12.26 Parachuting 229
- 12.27 Radioactive Waste at a Nuclear Power Plant 230
- 12.28 Visualizing Euler's Method 231
- 12.29 The Brightest Phase of Venus 233

A The Maple Interfaces — 235
- A.1 Selecting an Interface and Mode 235
- A.2 The Classic Interface 236
- A.3 The Standard Interface 237
- A.4 The Maplesoft Website 240

Index — 241

Introduction

Maple is a powerful software tool for mathematical computations and visualization. Maple (and other computer algebra systems) have radically changed the way scientists and engineers do mathematics in the same way that calculators changed computation in the 1970's. The goal of this manual is to introduce Maple to students who are taking first year calculus. As such, Maple is a tool to solve problems that are too difficult to solve by hand. In addition, we hope that by using Maple to solve problems, students will also improve their understanding of the concepts of calculus. The order of the material is organized by computational topic and should be suitable for most texts on single variable calculus.

This manual is written for a calculus course that involves at least one hour in a computer lab per week (with some additional hours outside of class). Since the students are new to Maple, the manual uses relatively few Maple commands to keep the syntax to a minimum. Most of the examples and exercises involve little formal programming with Maple. As another disclaimer, this manual will not magically elucidate all the concepts of calculus (although we hope it will help). Students of calculus must still rely on good lectures by instructors (unless the class size is small enough to allow a more radical approach) and still do lots of homework problems (some with Maple and many without).

The first part of this manual (Chapters 1 - 9) reads like a standard text, introducing Maple by example. Chapters 1, 2 and 3 introduce basic data types, including functions, expressions, equations, data sets and parametric curves, and acquaint students with some essential Maple commands to manipulate these, including evaluating, solving, plotting and curve fitting. Chapters 4 through 9 cover the main topics of single variable calculus: limits, derivatives, integrals, differential equations and series. In addition there are Sections 3.1 on Curve Fitting, 5.4 on Designer Polynomials and 7.4 on Fourier Cosine Series which go beyond the standard calculus material. Exercises are given at the end of each chapter.

The second part of this manual (Chapters 10 - 12) contains an introduction to Maple programming, trouble shooting tips and 29 projects. Chapter 10 describes Maple's conditional, looping and procedural structures, i.e. `if` statements, `do` loops and `proc` commands. Chapter 11 discusses the top 10 problems encountered by new Maple users. Students are advised to read through Chapter 11 as they read Chapters 1 - 3. Chapter 12 contains 29 student projects of varying levels of difficulty. Most of the projects apply Maple in the problem-solving process to simplify computations and to help with graphics. The projects are generally ordered according to calculus topic required and the level of difficulty. The beginning of each project contains a brief statement on the concepts from calculus that are required. Hints and guides as to how Maple can be utilized are sometimes given (especially for the earlier projects).

The examples, exercises and projects in this manual vary considerably in length and difficulty. Some are routine and designed to illustrate Maple syntax. Others are more involved and are designed to solve problems or illustrate ideas that would be difficult using only hand computation. Many of the more complicated examples and exercises are designed to embellish classical problems with more real-life considerations. For example, consider the problem of minimizing the surface area of a cylindrical can of fixed volume. Most calculus books in print over the past several generations contain this problem as an example or an exercise. After this problem has been translated into mathematics, its solution is straightforward and does not require the use of a computer. However, the computer allows the student to consider a more real-life version of this problem (See Section 5.6 and Exercise 11.) that seeks to minimize the cost of constructing this can where the cost of the seam used to attach the top, bottom and sides is considered along with the cost of the material. Minimizing this cost function requires the use of the computer for computations and graphics. Students should be expected to do the classical version of this problem by hand, and then they can be assigned the more real-life version of this problem in the computer lab.

As a word of warning to students, Maple alone will not solve calculus problems. "Thinking" is required for problem set-up and solution interpretation. Maple can only help with the computations and graphics that are necessary to obtain final answers. With this in mind, students should read the text of each chapter and set up any assigned problems before coming to their computer lab in order to make the best use of time spent in front of the computer.

Maple 10 through 12 come with two worksheet interfaces, a Classic Interface and a Standard Interface written in Java which is slightly slower but has several enhancements such as improvements to the plotting interface and the help system and direct access to many Maplet tutorials. For more information on the choice of interface and options, see Appendix A. Everything in this book has been tested with Maple 10 through 12 using both the Classic Interface and the Worksheet Mode of the Standard Interface. Most of it should also work with Maple 7, 8, 9 or 9.5. All of the commands should work the same on all windowing operating systems: (*Windows, Unix, Linux* or *Macintosh*).

INTRODUCTION

Historical Information: This manual was originally written in the spring of 1995 with Maple V Release 3 and was independent of any particular textbook. Subsequent editions were modified to match various editions of the Calculus textbooks by James Stewart. The current edition is again independent of any particular textbook. In each edition, the files were originally written as Maple worksheets and then exported as LaTeX files in order to ensure that no errors (hopefully) were made in transcribing Maple output into the text. The various editions were:

CalcLabs with Maple V,
1995, (for Maple V Release 3) ISBN: 0-534-25590-6

CalcLabs with Maple for Stewart's Calculus Concepts and Contexts, Single Variable,
1997, (for Maple V Release 4) ISBN: 0-534-34442-9

CalcLabs with Maple for Stewart's Single Variable Calculus Fourth Edition,
1999, (for Maple V Release 5) ISBN: 0-534-36433-0.

CalcLabs with Maple for Stewart's Single Variable Calculus Concepts and Contexts, Second Edition,
2001, (for Maple V Release 4) ISBN: 0-534-37922-2.

CalcLabs with Maple for Stewart's Single Variable Calculus Fifth Edition,
2003, (for Maple 8) ISBN: 0-534-39370-5.

CalcLabs with Maple for Stewart's Single Variable Calculus Concepts and Contexts, Third Edition,
2005, (for Maple 9.5) ISBN: 0-534-41026-X.

CalcLabs with Maple for Stewart's Single Variable Calculus Sixth Edition,
2008, (for Maple 10 and 11) ISBN: 978-0-495-01235-1.

CalcLabs with Maple Commands for Single Variable Calculus,
2009, (for Maple 12) this book.

All author royalties on copies of this manual sold to Texas A&M University students have been donated to an undergraduate scholarship fund for Texas A&M University students.

Please send all comments and corrections to:

Philip Yasskin, Department of Mathematics, Texas A&M University, College Station, TX 77843-3368 or yasskin@math.tamu.edu

Syllabi for Calculus I and II

The following syllabi assume that students have one hour per week of organized computer laboratory instruction together with two or three hours per week of outside access to a computer during a fourteen week semester. In class, the students work in a self-paced mode, alone or in groups, using the computer to try the examples in the text and solve assigned exercises or projects. The lab instructor should circulate to answer questions or provide guidance on troublesome syntax issues (such as the distinction between expressions and functions). Outside class, students should read the assigned chapters to prepare for lab and must complete the assigned exercises or projects by the due date.

For each weekly lab, some instructors prefer to assign 4 or 5 exercises from the end of Chapters 1 – 9. Others assign a shorter project from Chapter 12. For these weekly labs, students should work in pairs. If they do not complete the assignment during the lab, they can complete it at home and turn it in at the beginning of the next lab. A few times during the semester, the students can be assigned a longer project from Chapter 12. On the projects, students should work in groups of 3 or 4 and have from 2 to 4 weeks to complete it. Some instructors prefer to assign the same projects to the whole class. Other instructors assign different projects to each group.

Calculus I Syllabus

Any of the first 11 projects from chapter 12 can be done with standard Calculus I tools. However, we advise caution here. Do not get too ambitious during the first semester since students need to adjust to Calculus and Maple syntax (and to all the other life-changing attributes of college life). The assignment of one to three projects is probably sufficient for those who plan to incorporate projects into Calculus I. The more ambitious projects can be postponed to Calculus II.

Calculus II Syllabus

At this point, the students should have a small level of Maple sophistication. Consequently, the instructor can assign some reading from chapter 10. The instructor can also assign one or two of the remaining projects in chapter 12.

Basic Maple Commands

- All Maple commands must be terminated with a semicolon (if output is desired) or a colon (to suppress output).

- A command is "**executed**" (and assignments are put into memory) by pressing the ⟨ENTER⟩ key. The whole worksheet is executed by clicking on the *!!!* icon on the tool bar.

- If Maple takes a very long time to execute a command, you can interrupt the computation by clicking on the red STOP sign on the tool bar.

- Help on the syntax of any Maple command can be obtained by typing *?command*. For example, to get help with the `solve` command, type `?solve`.

- See Chapter 11, for many tips to help you avoid some of the typical problems with Maple.

A portion of the Maple commands used in this manual are listed below:

`a:=1.53;`   Assigns the value 1.53 to the variable a.
`a:=Pi*r^2;`   Assigns the formula πr^2 to the variable a.
`a:='a';`   Unassigns any value previously given to a.
`%`   Refers to the output of the previous command.
`f:=x^2+5;`   Defines the *expression* $f = x^2 + 5$.
`f:=x->x^2+5;`   Defines the *function* $f(x) = x^2 + 5$.
`eqn:=x^2-3=2*x;`   Assigns the *equation* $x^2 - 3 = 2x$ to the variable *eqn*.
`lhs(eqn);` and `rhs(eqn);`   Read off the left and right hand sides of an equation.
`Pi`   The *exact* constant π.
`exp(x)`   The exponential function, e^x.
`exp(1)`   The *exact* constant e.
`evalf(expr);`   Evaluates `expr` as a floating point decimal number.
`expand(x^2*(2*x+1)^3);`   Distributes multiplication over addition.
`expand(sin(x+y));`   Uses the formula for the sine of the sum of two angles to get $\sin(x)\cos(y) + \sin(y)\cos(x)$.
`expand(exp(a+b));`   Expands e^{a+b} to the product $e^a e^b$.

`simplify(expr);` Algebraically simplifies `expr`.

`factor(expr);` Factors a polynomial `expr`.

`eval(expr, x=a);` Substitutes a for x at each occurrence of x in `expr` and simplifies the result. The value of a could be a number or an algebraic expression like $2y + 3$.

`subs(x=a,expr);` Substitutes a for x at each occurrence of x in `expr`. The value of a could be a number or an algebraic expression like $2y + 3$.

CAUTION: The command `subs(x=a,expr);` *only* substitutes $x = a$ into the expression *expr*. It does *not* simplify and does *not* change any previously assigned value of x.

`solve(eqn,x);` Tries to give an *exact* solution listing all x's which solve an equation. If the equation is a polynomial of degree three or more, the solution may be expressed as `RootOf`'s which may represent multiple solutions. To separate the solutions, use `allvalues(%)`. Decimal approximations may be obtained using `evalf(%)`. Maple may not know how to find *exact* solutions in some cases.

`fsolve(eqn,x);` Tries to find *approximate floating point decimal* solutions of an equation. If the equation is polynomial `fsolve` will return all real solutions. If the equation is non-polynomial `fsolve` will return one real solutions. In that case, use the `avoid` parameter or a range to find each other solution.

`fsolve(eqn, x=a..b, avoid={x=c1, x=c2});` Tries to find *approximate* solutions of an equation in the interval $a \leq x \leq b$, avoiding the solutions at `x=c1` and `x=c2`. Use a plot to find a range for each solution.

`sol:=solve({eqn1,eqn2},{x,y});` Tries to find *exact* solutions x and y to the system of simultaneous equations labeled `eqn1` and `eqn2`. If there is more than one solution, they are `sol[1]`, `sol[2]`, etc.

`subs(sol,[x,y]);` Converts the values of x and y found above into an ordered pair.

`plot(expr,x);` Plots `expr` over the default interval $-10 \leq x \leq 10$. The scale of the y-axis is adjusted to fit the y values that appear in the plot. The scales for the x-axis and the y-axis need not match.

`plot(expr,x=a..b);` Plots `expr` over the interval $a \leq x \leq b$.

`plot(expr,x=a..b,y=c..d);` Plots `expr` over the interval $a \leq x \leq b$, but restricts the displayed values of y to the range $c \leq y \leq d$.

`plot(expr,x, scaling=constrained);` Plots `expr` over the default interval $-10 \leq x \leq 10$ using the maximum range of y's to determine the y-scale, but adjusting the x scale to correspond to the scale given for y so that there is less distortion.

`plot([expr1,expr2],x, color=[red,blue]);` Plots the graphs of `expr1` and `expr2` on the same coordinate axes with respective values of the options.

`plot(f,2..3);` Plots the *function* `f` over the interval $2 \leq x \leq 3$. Note that when a *function* is plotted, as opposed to an *expression*, no "x" is included in the plot command.

BASIC MAPLE COMMANDS

`with(plots):` Loads the `plots` package that is necessary for the `implicitplot` and `display` commands.

`implicitplot(x^2+y^2=1, x=-1..1, y=-1..1);` Plots the *equation* $x^2 + y^2 = 1$.

`plot([f(t),g(t),t=a..b]);` Plots the parametric equations $x = f(t)$ and $y = g(t)$ for $a \leq t \leq b$.

`display([p1,p2]);` Combines plots p1 and p2.

`Limit(expr,x=a);` Displays (but does not evaluate) the limit $\lim_{x \to a} expr$.

`value(%);` Evaluates the previous limit.

`limit(expr,x=a);` Computes the limit $\lim_{x \to a} expr$. It is better to use `Limit(expr,x=a)` and `value(%)` unless `expr` was previously displayed.

`limit(expr,x=infinity);` Evaluates the limit $\lim_{x \to \infty} expr$.

`Diff(expr,x);` Displays (but does not evaluate) the derivative $\frac{d}{dx} expr$ of the *expression* `expr`. The x is required even if there are no other variables in `expr`.

`value(%);` Evaluates the previous derivative.

`diff(expr,x);` Computes the derivative $\frac{d}{dx} expr$ of the *expression* `expr`. It is better to use `diff(expr,x)` and `value(%)` unless `expr` was previously displayed.

`D(f);` Returns the derivative of the *function* f as a *function*.

`Int(expr,x);` Displays (but does not evaluate) the indefinite integral $\int expr\, dx$. The x is required even if there are no other variables in `expr`.

`value(%);` Evaluates the previous integral.

`int(expr,x);` Computes the indefinite integral $\int expr\, dx$. The answer is another *expression*. It is better to use `Int(expr,x)` and `value(%)` unless `expr` was previously displayed.

CAUTION: Maple does *not* add a constant of integration C to indefinite integrals.

`Int(expr,x=a..b);` Displays (but does not evaluate) the definite integral $\int_a^b expr\, dx$.

`value(%);` Evaluates the previous integral.

`int(expr,x=a..b);` Computes the definite integral $\int_a^b expr\, dx$. It is better to use `Int(expr,x=a..b)` and `value(%)` unless `expr` was previously displayed.

`with(student):` Loads the `student` package, necessary for the `changevar` and `intparts` commands.

`changevar(3*x=tan(theta),A,theta);` Performs the change of variables $3x = \tan(\theta)$ in the integral labeled A.

`intparts(A,uexpr);` Performs one integration by parts on the integral labeled A with `uexpr` used as the u variable.

`dsolve(deq, y(x));` Solves the differential equation labeled `deq` for the unknown function $y(x)$. The constants of integration are denoted by `_C1`, etc.

`dsolve({deq, init}, y(x));` Solves the differential equation labeled `deq` with the initial condition labeled `init` for the unknown function $y(x)$.

`Sum(expr,i=m..n);` Displays (but does not evaluate) the sum $\sum_{i=m}^{n} expr$.

`Sum(expr,i=m..infinity);` Displays (but does not evaluate) the series $\sum_{i=m}^{\infty} expr$.

`value(%);` Evaluates the previous sum or series.

`sum(expr,i=m..infinity);` Computes the series $\sum_{i=m}^{\infty} expr$. It is better to use `Sum(expr,i=m..infinity)` and `value(%)` unless `expr` was previously displayed.

`p:=taylor(f,x=a,n);` Assigns to p the Taylor expansion of the expression f about the point $x = a$ with an error term of order n.

`p:=convert(p,polynom);` Converts p to a polynomial (removing the error term in a Taylor expansion).

Saving, Restarting, Deleting and Setting Options

- To save the worksheet, select FILE + SAVE or click on the save icon.

- When you reopen the worksheet, everything is on the screen but nothing is in memory. To re-execute the whole worksheet, select EDIT + EXECUTE + WORKSHEET or click on the *!!!* icon. Remember the worksheet is re-executed in order from top to bottom. So if you typed or modified your worksheet out of order, it may not re-execute properly.

- To clear out Maple's memory without closing and reopening Maple, execute `restart;` It is sometimes a good idea to execute a `restart` before starting each new problem.

- To delete a line of input or output or a plot, press ⟨CTRL-DELETE⟩.

- To delete all of the output, select EDIT + REMOVE OUTPUT + FROM WORKSHEET. It is a good idea to remove all of the output before saving. This reduces the size of saved files and reminds you that you need to re-execute the worksheet when you reopen Maple.

- See Appendix A for information on selecting the proper Maple interface and setting options including Maple's input mode and its AUTOSAVE feature.

Chapter 1

Getting Started

Before starting, please read Appendix A so that your Maple interface is properly configured. This chapter introduces some of the basic Maple commands associated with assigning labels to numbers and formulas and creating plots.

1.1 Maple as a Calculator

A Maple input line is indicated by an input prompt > at the left hand margin. A Maple statement (or command) is entered (or executed) by typing it on an input line with a semicolon (;) at the end and then pressing the ⟨ENTER⟩ or ⟨RETURN⟩ key. Try entering

> 2 + 5;
$$7$$

(Don't type the > prompt since this is provided by the computer.) Maple's output, 7, is displayed in the center of the next line.

If a command is entered and Maple replies

> 2 + 5

    Warning, premature end of input

then you have forgotten the semicolon. Click at the end of the line, type a semicolon and press ⟨ENTER⟩ again. (See Section 11.1.)

If a command or an expression is entered incorrectly then click the mouse on the line, edit it and then re-execute it by pressing ⟨ENTER⟩ again.

Maple can do arithmetic on formulas entered on an input line. The standard arithmetic operations are

+ addition   − subtraction
* multiplication   / division
^ exponentiation

The standard order of operations is exponentiation before multiplication and division and then addition and subtraction. Within a level, commands are done left to right. To be safe, use parentheses to be sure that the operations are performed in the desired order. For example, (3+4)/7; is not the same as 3+4/7;. Be sure to use round parentheses ( ) rather than square brackets [ ] or curly braces { }, which have other meanings in Maple.

CAUTION: A typical mistake is typing

> (2+4)(3-1);

$$6$$

instead of

> (2+4)*(3-1);

$$12$$

Maple will not multiply without the * sign. Rather it will give very peculiar results without warning. Try it! (See Section 11.1.)

Maple knows a large number of standard mathematical functions including:

square root	sqrt
absolute value	abs
natural exponential	exp
natural logarithm	ln
trig functions	sin, cos, tan, sec, csc, cot
inverse trig functions	arcsin, arccos, arctan, arcsec, arccsc, arccot

It is important to understand the distinction between numbers that Maple knows exactly, such as 2, 1/3, $\sqrt{2}$, and π, and *floating point decimal numbers* such as 2.0, .33333, 1.414, and 3.14. The number 1/3 is an expression that represents the exact value of one-third, whereas .33333 is a floating-point decimal approximation of 1/3.

If numbers are entered as integers, then Maple normally returns exact answers. For example, enter

> (1+3)/6;

$$\frac{2}{3}$$

Maple returns 2/3 rather than its decimal approximation .6666666667. To get the decimal approximation, use the `evalf` command. For example, try

> evalf(22/79+34/23);

$$1.756741882$$

Notice that parentheses are needed around the expression. Here, `evalf` means to evaluate the expression as a floating-point decimal number. Alternatively, one of the numbers can be entered as a decimal. For example, type

> 22./79 + 34/23;

$$1.756741883$$

and Maple returns the answer in decimal form.

1.2 Assigning Variables

Maple answers are often used again in subsequent calculations and therefore Maple provides a way to store and recall earlier results. One way to refer to an earlier result is to use the percent sign %, which refers to the immediately preceding result. For example, the calculation of $2^6 + 1$ can be done in two steps to show intermediate answers. The result of the first operation is given to the second part as %.

> `2^6; %+1;`

$$64$$
$$65$$

Note that the input line contains two Maple statements (since there are two semicolons) and therefore there are two Maple outputs.

Names or labels can also be used to store and refer to results. For example, the number $22./79 + 34/23$ can be assigned to the variable a by typing

> `a:=22./79+34/23;`

$$a := 1.756741883$$

Maple commands of this form are called assignment statements and the := sign indicates that the quantity on the right is to be assigned to the variable name on the left. Now, the number 1.756741883 can be recalled by typing a.

EXAMPLE 1: Compute $(1.756741883)^2$, $\dfrac{1}{1.756741883}$, and $\sqrt{1.756741883}$.

> `a^2;`

$$3.086142043$$

> `1/a;`

$$0.5692355887$$

> `sqrt(a);`

$$1.325421398$$

To more easily distinguish between various labels, use descriptive names.

EXAMPLE 2: Enter an expression that describes the area of a circle of radius r

> `Area:=Pi*r^2;`

$$Area := \pi r^2$$

Note that π is entered with an upper case P. With a lower case p, Maple will show the Greek letter π but won't recognize its mathematical meaning. To evaluate this area when $r = 5$, enter `r:=5;`. The value $r = 5$ will automatically be substituted into *Area*.

> `r:=5; Area; evalf(%);`

$$r := 5$$
$$25\pi$$
$$78.53981635$$

Note, however,

> `evalf(pi*r^2);`

25. π

has the wrong type of π. Also note that labels are case sensitive: the label `Area` is different from the label `area`.

EXAMPLE 3: Compute the profit on an item if its retail price is $4.95 and its wholesale cost is $2.80.

SOLUTION: Enter the price and cost:

> `Price:=4.95;`

$$Price := 4.95$$

> `Cost:=2.80;`

$$Cost := 2.80$$

Then the profit is given by

> `Profit:=Price-Cost;`

$$Profit := 2.15$$

NOTE: A variable keeps its value until it is assigned a new value or until it is cleared (or unassigned) or until Maple is restarted. For example, the value of the variable `Price` can be unassigned by assigning it to its name in single quotes:

> `Price:='Price';`

$$Price := Price$$

Alternatively, the values of the variables `Cost` and `Profit` can be unassigned by using the `unassign` command:

> `unassign('Cost','Profit');`

Now the variables `Price`, `Cost` and `Profit` have no values assigned to them.

More generally, the command

> `restart;`

will clear out everything from Maple's memory, thus unassigning all variables.

TIP: In fact it is probably useful to execute a `restart` at the beginning of each new homework problem.

The same thing happens if you save your worksheet (by selecting FILE > SAVE or clicking on the save icon) and reopen it at a later date. Everything is on the screen but nothing is in memory. To re-execute the whole worksheet, select EDIT > EXECUTE > WORKSHEET or click on the *!!!* icon. Remember the worksheet is re-executed in order from top to bottom. So if you typed or modified your worksheet out of order, it may not re-execute properly.

1.3 Algebra Commands

We have seen how to manipulate numbers and assign them to variables (or labels). Maple can also manipulate algebraic expressions involving variables.

For example, to multiply out the expression $(3x-2)^2(x^3+2x)$, type

> `(3*x-2)^2*(x^3+2*x); expand(%);`

1.3. ALGEBRA COMMANDS

$$(3x-2)^2 (x^3 + 2x)$$
$$9x^5 + 22x^3 - 12x^4 - 24x^2 + 8x$$

As mentioned earlier, the percent % refers to the output preceding the percent, in this case, the expression $(3x-2)^2(x^3+2x)$.

NOTE: It is easy to make a typing error when entering a complicated expression, such as $(3x-2)^2(x^3+2x)$. To prevent such errors from affecting a Maple command (such as expand), first type the expression without the command and press ⟨ENTER⟩ as follows:

> (3*x-2)^2*(x^3+2*x);

$$(3x-2)^2 (x^3 + 2x)$$

Examine Maple's output to make sure that the expression is entered correctly. Then click the mouse back at the end of the previous line and add the Maple command expand(%) as done above. Putting expand(%) on the same line also guarantees that if you change the expression to expand, the expand(%) will also be re-executed.

To factor the polynomial $x^6 - 1$, type

> x^6-1; factor(%);

$$x^6 - 1$$
$$(x-1)(x+1)(x^2+x+1)(x^2-x+1)$$

Another useful command is simplify. For example, to simplify the expression $\dfrac{x^2-x}{x^3-x} - \dfrac{x^2-1}{x^2+x}$ enter

> (x^2-x)/(x^3-x)-(x^2-1)/(x^2+x); simplify(%);

$$\frac{x^2 - x}{x^3 - x} - \frac{x^2 - 1}{x^2 + x}$$
$$-\frac{x^2 - x - 1}{x(x+1)}$$

The following command will also simplify the expression.

> simplify((x^2-x)/(x^3-x)-(x^2-1)/(x^2+x));

$$-\frac{x^2 - x - 1}{x(x+1)}$$

However, with this syntax, it is harder to keep track of the parentheses in such a long expression. In addition, this command does not display the original expression and therefore it cannot be checked for typing errors.

NOTE: Maple has an on-line Help facility that is invoked by typing ? followed by the command. For example, to get help with the factor command, type (No semicolon is necessary.)

> ?factor

Alternatively, if the command word is already typed in the worksheet, just click on it and press F2. Also explore the Help Browser and the Search facilities by clicking on HELP > MAPLE HELP.

1.4 Plots

The `plot` command is best introduced with an example. To plot the graph of

$$f = \frac{x^2 - 4}{x + 1}$$

over the interval $-6 \leq x \leq 6$, type

```
>  f:=(x^2-4)/(x+1);
```

$$f := \frac{x^2 - 4}{x + 1}$$

```
>  plot(f, x=-6..6);
```

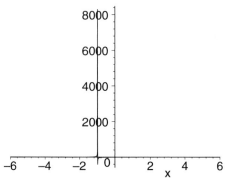

The scale on the y-axis is much different from the scale on the x-axis because of the very large function values when x is close to -1 (where the function becomes undefined). To get a more reasonable plot, the y-range should be specified. For example, to view the piece of the graph with $-10 \leq y \leq 10$, enter

```
>  plot(f, x=-6..6, y=-10..10);
```

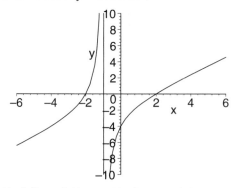

Notice the vertical line at the vertical asymptote $x = -1$. This is an artifact of the way Maple makes plots. It calculates points and connects them with straight lines. To see these points, right-click in the plot and select STYLE > POINT. Try that above. Alternatively, include the plot option `style=point` as follows:

1.4. PLOTS

> `plot(f, x=-6..6, y=-10..10, style=point);`

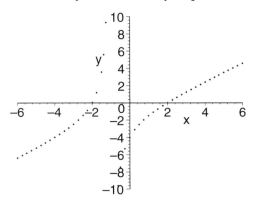

By default, Maple plots 49 points. To increase the number of points, add the option `numpoints=#` (but don't make the number absurdly large):

> `plot(f, x=-6..6, y=-10..10, style=point, numpoints=201);`

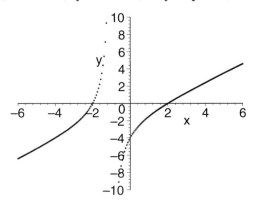

To connect the dots, while eliminating the vertical line, include the option `discont=true`:

> `plot(f, x=-6..6, y=-10..10, discont=true);`

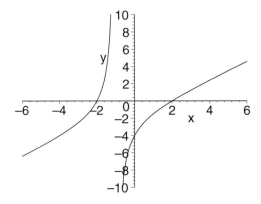

Next notice that the scale on the y-axis is different from the scale on the x-axis. Normally, Maple adjusts the scales of both axes so that the plot fills the plot window. To equalize scales on the x- and y-axes, right-click in the plot and select SCALING CONSTRAINED. Try that above. Alternatively, include the option `scaling=constrained`

> `plot(f, x=-6..6, y=-10..10, discont=true, scaling=constrained);`

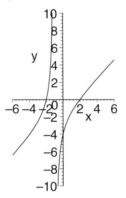

Other options to `plot` will be discussed in Chapter 2. A complete list of the plot options may be seen by executing

> `?plot,options`

By changing the plot range, different aspects of the graph can be viewed. For example, the above plot shows the x-intercepts at -2 and 2, the y-intercept at -4 and the vertical asymptote at $x = -1$. Changing the x- and y-ranges to `x=-200..200, y=-200..200` displays the graph for larger values of x:

> `plot(f, x=-200..200, y=-200..200, discont=true);`

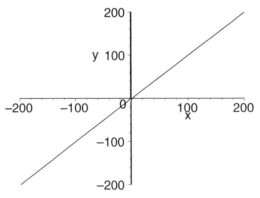

Notice the graph of the function approaches the line $y = x - 1$ which is a slant asymptote. (You will learn later how to compute slant asymptotes.) However, with such large values of x, the vertical asymptote at $x = -1$ becomes obscured.

If the x-range is omitted from the plot command, Maple will plot the expression over the interval $-10 \leq x \leq 10$. (In other words, this is the default range).

More than one expression can be graphed on the same plot by enclosing several expressions using curly braces { } or square brackets [ ]. For example, to add the skewed asymptote $y = x - 1$ to the above plot, type

> `g:=x-1;`

$$g := x - 1$$

> `plot({f,g}, x=-6..6, y=-10..10, discont=true);`

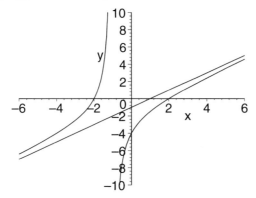

CAUTION: Opening too many plots on your computer may degrade its performance. It is wise to close unneeded plots by clicking on the plot and pressing ⟨CTRL-DELETE⟩. Likewise, to reduce the size of saved files, click on the EDIT menu and select REMOVE OUTPUT > FROM WORKSHEET before saving. To recover the output, click on the EDIT menu and select EXECUTE > WORKSHEET or click on the toolbar icon which looks like three exclamation points, *!!!*.

1.5 Summary

- Maple does everything that a graphing calculator will do.

- Maple function arguments must be in parentheses; for example, $\sin x$ must be written as `sin(x)`.

- Maple executes arithmetic commands in a predefined order of precedence. Use parentheses to modify the order. Use care in entering expressions and examine output for correctness.

- Multiplication requires an asterisk. Juxtaposition of symbols is not allowed in Maple as a synonym for multiplication: `(x+4)(x+2)` is viewed as a function evaluation, not a product.

- Most calculators store numbers only as floating-point decimals, with a preselected number of digits of accuracy. Maple has an alternative exact mode that treats the fraction $1/3$ not as a decimal 0.3333333333 to any number of 3's, but instead as a list of two integers, 1 and 3.

- The distinction between storage modes is important because Maple will often not be able to express an answer in terms of a list of integers, and so it will parrot back the original expression typed in. This does not represent a syntax error: Maple simply doesn't know how to give an exact answer for `sin(1)`.

- Exact answers can be converted to floating-point decimal approximations of any number of digits by using `evalf`.

- It is unnecessary (and unwise) to retype intermediate results in subsequent calculations or even to copy and paste them. It is unwise because if you correct a previous mistake or even just change something, the copied result will not automatically update. Maple provides two alternatives: the assignment command, `:=`, in which a label is associated with a given output; and the percent, `%`.

- Once a number or algebraic expression is assigned a label, any statement that contains that label treats it as a synonym for the number or expression itself.

- Once an assignment has been made, Maple remembers that assignment until it is told otherwise. A label may be unassigned by reassigning it to its own name enclosed in single forward quotes or by including it in an `unassign` command or by executing a `restart`. One common source of frustration is forgetting that a label already has an assigned value when trying to use it as a free variable. (See Section 11.4.)

- Save your worksheet frequently. When you reopen a worksheet, re-execute it by clicking on the *!!!* icon.

- Algebraic expressions can be manipulated with the commands `expand`, `factor`, and `simplify`.

- Use `?command` to get help on a command including proper syntax and examples. The command name does not have to be exact to get an answer. Also, use the Help Browser and Search facility under the HELP menu.

- Maple can be used to create graphs of expressions. Know how to use Maple to plot a graph, how to modify the domain and range, and how to specify options. Be aware that the defaults do not always show important features of the graph and know how to change the defaults to do so.

- Clicking the mouse on a graph shows the approximate coordinates of the point on the top left corner of the Maple window.

1.6 Exercises

1. Assign the variable name a to the number $2\pi/12$. Evaluate each of the following exactly and then use `evalf(%)` to compute the decimal approx-

imations:
a^2, $1/a$, $\sqrt{a}$, $a^{1.3}$, $\sin(a)$, $\tan(a)$, and $\tan^{-1}(a)$.

2. The number of significant digits can be temporarily changed from the default value of 10 by modifying the `evalf` command. For example, to calculate a^2 to 20 significant digits, type `evalf(a^2,20);` Repeat Exercise 1 with 20 significant digits using `evalf(%,20)`.

3. The number of significant digits can be permanently changed from the default value of 10 to some other number, such as 25, with the command `Digits:=25`, Repeat Exercise 1 with 25 significant digits. Remember to return the number of digits to 10 by executing `Digits:=10;`.

4. Display the following expressions and then `expand` them.

    (a) $(x^2 - 4x + 3)^3(x^2 + 4)$
    (b) $(x - a)^6$

    NOTE: If you have done Exercise 1, 2 or 3, the label a already has a value assigned to it. Recall that this value should be unassigned by typing `a:='a';` before part b..

5. Display and `factor` the expression $x^2 - \frac{5}{3}x - \frac{2}{3}$. What happens if this expression is changed to $x^2 - \frac{5.0}{3}x - \frac{2.0}{3}$?

6. Try factoring $x^2 - 4x - 9$. What happens? Repeat with the command `factor(%,sqrt(13))`.

7. Factor $x^8 - 1$ (over the integers). Repeat with each of the following as the second argument of `factor`, and discuss the differences.

    (a) `sqrt(2)`
    (b) `I`
    (c) `{sqrt(2),I}`
    (d) `real`
    (e) `complex`

8. Simplify
$$\frac{8x^2}{x^4 - 1} - \frac{4}{x^2 - 1}$$
What is the difference in the domains of the original expression and its simplified form?

9. Plot the graph of $\sec(x)$. Experiment with the x- and y-ranges to obtain a reasonable plot of one period of $\sec(x)$. Eliminate the vertical asymptote.

10. (a) Plot the graph of $\dfrac{4x^2 - 2x + 2}{x - 1}$ over a small interval containing $x = 1$, for example, $0 \leq x \leq 2$. Experiment with the y-range to obtain a reasonable plot. What happens to the graph near $x = 1$? Add the option `discont=true`.

    (b) Now plot the same expression over a large interval such as $-100 \leq x \leq 100$ without `discont=true`. Note that the behavior of the graph near $x = 1$ is no longer apparent. Why do you think this happens? Try adding `style=point`. Now add `discont=true`, both with and without `style=point`.

11. Repeat Exercise 10a for $\dfrac{4x^2 - 2x - 2}{x - 1}$. To explain the result try simplifying the expression. (This is an example of a function which is undefined at $x = 1$ but whose limit is defined there.)

12. Plot the expressions $\sin(x)$, $\sin(2x)$, and $\sin(4x)$ over the interval $0 \leq x \leq 2\pi$ on the same coordinate axes. Now plot the same expressions over the interval $0 \leq x \leq 4\pi$. At what numbers are they all equal?

13. Use `evalf(Pi,40);` to give the first 40 digits of π.

14. Compute the exact and floating-point values of $\sin(\pi/3)$ and $\sin(3)$.

15. Compute the number of seconds in one year, showing the units in your product as each factor is entered, e.g. `365*day/year`.

16. Just as `factor` will factor a polynomial, there is also an `ifactor` (integer factor) command that gives the prime decomposition of an integer. Use `ifactor` to show that $2^{21} - 1$ is not prime. How long would it have taken you to find the factors by hand?

17. Compute 27!. Recall: $4! = 4 * 3 * 2 * 1$. How many final zeros are there? Factor 27! using `ifactor`. How many factors of 5 are there? Why is this not a coincidence? NOTE: Maple knows the ! sign.

18. What happens when the command `expand` is applied to $(a - 2b)/c$? Practice the technique of entering the expression and checking its Maple output to see that it is entered correctly. Then go back and add `expand(%)` on the same line.

19. Apply `expand` to $\ln\left(\dfrac{a}{bc}\right)$. Repeat this but preceded by `assume(a>0, b>0, c>0);` which tells Maple a, b and c are positive. Why did this make a difference? To remove the assumptions, execute `a:='a';` and similarly for b and c. NOTE: The tildes (~) mean there are assumptions on the variables. See Appendix A to eliminate the tildes.

20. Factor the expression $e^{2x} - 1$, by first using `expand`, then `factor`. Compare this with just using `factor`.

21. Load the `student` package by typing `with(student);` Label the points $P_0 := [1, 3]$ and $P_1 := [4, 7]$ on the same input line. (NOTE: Subscripts are entered by putting them in square brackets, e.g. `P[0]`.) Find out the syntax for determining the slope between the two points via `?slope`. Find the slope between P_0 and P_1 using the `slope` command. Assign the value that you find to the variable m. Find the equation of the line through P_0 and P_1.

22. Plot the expressions $-x/2 + 3/2$ and $-3x + 4$ on the same graph with x-values between 0 and 2. Right click in the plot and select SCALING CONSTRAINED to avoid distortion in the plot. Click on the intersection with the mouse to find the coordinates of the intersection of the lines $y = -x/2 + 5/2$ and $y = -3x + 5$. The coordinates are shown at the top-left of the Maple window just under the FILE menu.

Chapter 2

Expressions, Functions and Equations

This chapter discusses the difference between expressions, functions and equations, and how they are defined, manipulated and plotted in Maple. It also describes how to solve one or more equations. Briefly:

- An **expression** is simply a formula for computing something. It may involve zero, one or more variables, arithmetic operations and functions, as listed in Section 1.1. For example, (See Section 2.1.) the area of a circle of radius r is given by πr^2 and entered as the Maple expression:
    > `Area:=Pi*r^2;`

- A **function** is a rule for computing an output for each given input. It is usually specified *explicitly* by saying that the output (the dependent variable) is given by a some expression involving the input (the independent variable). For example, (See Section 2.1.) the area A of a circle is a function of its radius r given by the rule $A(r) = \pi r^2$ and entered as the Maple (arrow-defined) function:
    > `A:=r->Pi*r^2;`

    However, a function may also be specified *implicitly* as the solution of some equation. For example, (See Section 4.4.) the cube root function is defined as the solution (for y) of the equation $x = y^3$.

- Finally, an **equation** is a statement that two expressions are equal to each other. For example, (See Section 2.3.) the statement that the area of a circle is 40 cm^2 is given by the equation $\pi r^2 = 40$ and entered into Maple as:
    > `eq:=Pi*r^2=40;`

    NOTE: The Maple statement `Area:=Pi*r^2;` is *not* an equation because there is an assignment symbol `:=` but no equal sign.

The distinction between expressions, functions and equations may seem pedantic, but mathematics makes this distinction and Maple handles them very differently.

When you complete this chapter, please preview Chapter 11. This may help you avoid some of the typical problems with Maple.

2.1 Expressions and Functions: Definition and Use

Definition and Evaluation of Expressions and Functions: In the previous chapter, we saw that we can assign labels to numbers (such as `Area:=25*Pi;`). In a similar way, labels can be assigned to expressions, functions and equations, (all of which may contain variables which may or may not have been previously assigned values). This is useful when repeatedly referring to a complicated formula. For example, the area of a circle is the expression

> `Area := Pi*r^2;`

$$Area := \pi r^2$$

You can find the area for a particular value of r (say $r = 5$) by substituting the value into the area using either the `subs` command or the `eval` command:

> `subs(r=5,Area); eval(Area,r=5);`

$$25\pi$$
$$25\pi$$

The difference is that `eval` simplifies the answer while `subs` does not:

> `subs(x=Pi/4,sin(x)); eval(sin(x),x=Pi/4);`

$$\sin(\frac{\pi}{4})$$
$$\frac{\sqrt{2}}{2}$$

The area of a circle can also be thought of as the function

> `A := r -> Pi*r^2;`

$$A := r \to \pi r^2$$

In this "arrow notation" for a function, r is the independent variable and πr^2 is the expression which gives the rule by which to compute the value of the dependent variable A. (Notice that the arrow is typed as a minus and a greater than sign (->) with *no space!*) For a given value of r, the corresponding value of A is denoted by $A(r)$. So if $r = 5$, then $A(5) = 25\pi$. In Maple we compute

> `A(5);`

$$25\pi$$

Notice that

> `A(r);`

$$\pi r^2$$

is just the *expression* for the area.

CAUTION: Above, we said you substitute numbers into expressions by using the `subs` command. There is another way to do it (which is not as good). You can simply assign a value to the variable, e.g.

> `r:=7;`
$$r := 7$$

Then this value is automatically substituted into any quantity that uses this variable, e.g.

> `Area;`
$$49\pi$$

The problem comes when you try to define another quantity which uses that variable. For example, the circumference of the circle is:

> `Circum:=2*Pi*r;`
$$Circum := 14\pi$$

The value is automatically substituted here too. If you want the formula with a variable `r`, then you must first unassign `r` by entering:

> `r:='r';`
$$r := r$$

(Those are forward single quotes (').) Now `r` is back to a free variable and `Area` is back to its formula in terms of `r`:

> `Area;`
$$\pi r^2$$

However, quantities which were entered while `r` had a value, retain that value, e.g.

> `Circum;`
$$14\pi$$

This is probably not what you wanted. This means you need to define (or redefine) the quantity while `r` is a free variable:

> `Circum:=2*Pi*r;`
$$Circum := 2\pi r$$

The moral is that it is best never to assign values to the free variables in your formulas; use `subs` or `eval` instead.

Expressions, functions and equations can involve more than one variable. For example, the volume of a cylinder of radius r and height h is the expression

> `Volume:=Pi*r^2*h;`
$$Volume := \pi r^2 h$$

or the function

> `V:=(r,h) -> Pi*r^2*h;`
$$V := (r, h) \to \pi r^2 h$$

In this case, the volume involves two variables r and h.

2.1. EXPRESSIONS AND FUNCTIONS: DEFINITION AND USE

Multiple substitutions can be entered at the same time. If simultaneous substitution is desired, then curly braces { } or square brackets [ ] must be used. For example, to substitute $r = 2$ and $h = 5$ into the expression for the volume of a cylinder, enter

> `subs({r=2,h=5},Volume);`

$$20\pi$$

If you don't include the braces, the `subs` command makes the substitutions one at a time from left to right. For example, in the following

> `subs(r=h,h=5,Volume);`

$$125\pi$$

`r` is first replaced by `h`, giving πh^3, and then `h` is replaced by 5 giving 125π. However, in the following `r` is replaced by `h` and `h` is simultaneously replaced by 5:

> `subs({r=h,h=5},Volume);`

$$5\pi h^2$$

More realistic examples of simultaneous and sequential substitutions occur in Example 1 of Section 3.3 and Example 2 of Section 3.4.

If you use `eval` then only simultaneous substitution is possible:

> `eval(Volume,{r=h,h=5});`

$$5\pi h^2$$

Thus `subs` gives you more control than `eval`.

Functions of two or more variables are evaluated using the usual function notation:

> `V(2,5);`

$$20\pi$$

Again notice that the *expression* for the volume can be recovered by applying the function to its variables:

> `V(r,h);`

$$\pi r^2 h$$

Using Expressions and Functions in Commands: Some commands, like `expand` and `factor`, can only be applied to expressions. Others, like composition, denoted by `@`, can only be applied to functions. Still others, like `plot`, can be applied to either but the syntax is different. For example, the expression

> `g:=2*x^3-5*x^2+x+2;`

$$g := 2x^3 - 5x^2 + x + 2$$

can be factored by executing

> `factor(g);`

$$(x-1)(x-2)(2x+1)$$

However, the arrow-defined function

> `f:= x -> 2*x^3-5*x^2+x+2;`

$$f := x \to 2x^3 - 5x^2 + x + 2$$

can only be factored after you evaluate at x to produce the expression f(x):

> `factor(f(x));`

$$(x-1)(x-2)(2x+1)$$

On the other hand, you can compose the function f and the function sin, by typing f@sin. Its value is then

> `(f@sin)(x);`

$$2\sin(x)^3 - 5\sin(x)^2 + \sin(x) + 2$$

which is the same as

> `f(sin(x));`

$$2\sin(x)^3 - 5\sin(x)^2 + \sin(x) + 2$$

However, to compose the expression g and the expression sin(x) you must use the subs or eval command

> `subs(x=sin(x),g);`

$$2\sin(x)^3 - 5\sin(x)^2 + \sin(x) + 2$$

The @ notation is much closer to standard math notation.

In the next section, we will discuss the plot command which has a different syntax for expressions and functions.

It should also be noted that the subs command is purely a symbolic replacement. Thus

> `subs(x^2=2, x^4+x^2+x);`

$$x^4 + 2 + x$$

simply replaces all occurences of x^2 by 2. To substitute for all x's you must use

> `subs(x=sqrt(2), x^4+x^2+x);`

$$6 + \sqrt{2}$$

In addition, the subs command does *no* simplification. It must be followed by simplify(%), if necessary. For example

> `f:=(y^2-x)/x; subs(y=x,f); simplify(%);`

$$f := \frac{y^2 - x}{x}$$

$$\frac{x^2 - x}{x}$$

$$x - 1$$

Converting Between Expressions and Functions: Since some commands only work with either a function or an expression, it is important to be able to change back and forth between them.

To convert from a function to an expression is easy. As mentioned above, if f is a function, then f(x) is the expression given by *applying* f to the variable x.

2.1. EXPRESSIONS AND FUNCTIONS: DEFINITION AND USE

Conversely, to convert from an expression to a function, there are two options: The first method is simply to retype the formula (or copy and paste it) and insert the arrow notation `x->` as we did in going from `g` to `f` above. The problem with this method is that the formula for f may not appear in the worksheet. It may be a complicated result of a long computation.

The second method involves the `unapply` command. When you *apply* a function to its variable, you get an expression. Conversely, when you *unapply* an expression to its variable, you get a function. For example, since `g` is the expression

```
>   g;
```
$$2x^3 - 5x^2 + x + 2$$

then

```
>   h:=unapply(g,x);
```
$$h := x \to 2x^3 - 5x^2 + x + 2$$

is the corresponding function. The second entry in `unapply(g,x)` tells Maple that the variable in the function will be x. (There may be other variables in the expression besides x.) This method can be very helpful since you don't need to write out the expression for h.

CAUTION: The following assignment does not work!

```
>   h:= x -> g;
```
$$h := x \to g$$

because then

```
>   h(x);
```
$$2x^3 - 5x^2 + x + 2$$

but

```
>   h(2);
```
$$2x^3 - 5x^2 + x + 2$$

also. You must use `unapply`!

If the function is to have more than two variables, then the additional variables appear as additional arguments to `unapply`. For example, the volume of a cylinder can be defined as the expression

```
>   Volume:=Pi*r^2*h;
```
$$Volume := \pi r^2 h$$

and converted to a function using:

```
>   V:=unapply(Volume,r,h);
```
$$V := (r, h) \to \pi r^2 h$$

2.2 Expressions and Functions: Plots

This section explains how to plot expressions and functions with one or two variables using the `plot` and `plot3d` commands.
NOTE: These commands cannot plot *equations*. For that use the `implicitplot` and `implicitplot3d` commands discussed in Section 2.4.
Two-Dimensional Plots: To plot the expression

```
> g:=2*x^3-5*x^2+x+2;
```

$$g := 2x^3 - 5x^2 + x + 2$$

over the interval $-2 \leq x \leq 3$, use the `plot` command to execute

```
> plot(g,x=-2..3);
```

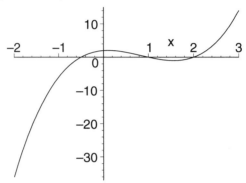

To plot the function

```
> f:=x->2*x^3-5*x^2+x+2;
```

$$f := x \to 2x^3 - 5x^2 + x + 2$$

over the same interval, execute

```
> plot(f,-2..3);
```

which gives the same plot. Note the difference in syntax. Since `f` is a function, it already knows its independent variable, so we don't need to (and must not) specify it with the interval. However, since `f(x)` is an expression, it is also valid to plot `f` using

```
> plot(f(x),x=-2..3);
```

which again gives the same plot. On the other hand, a mix of syntax will result in error messages or empty plots.

```
> plot(f,x=-2..3);

Error, (in plot) expected a range but received x = -2 .. 3

> plot(f(x),-2..3);

Warning, unable to evaluate the function to numeric values in the
region; see the plotting command's help page to ensure the calling
sequence is correct

Plotting error, empty plot
```

2.2. EXPRESSIONS AND FUNCTIONS: PLOTS

```
>  plot(g,-2..3);
```
```
Warning, unable to evaluate the function to numeric values in the
region; see the plotting command's help page to ensure the calling
sequence is correct
```
```
Plotting error, empty plot
```

To graph two or more expressions or functions in the same plot with the same interval, enclose them in curly braces { } or square brackets [ ]. For example, to graph g (from above) and

```
>  h:=2*x;
```
$$h := 2\,x$$

execute

```
>  plot({g,h},x=-2..3);
```

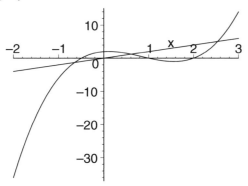

If you enclose them in square brackets, then options (like color, linestyle and thickness) can be specified separately for each expression:

```
>  plot([g,h],x=-2..3, linestyle=[SOLID,DOT], thickness=[3,1]);
```

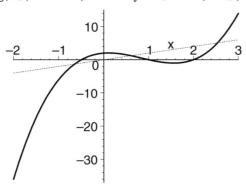

NOTE: The two expressions must have the same variable and range. On the other hand, functions may have different variables. For example, to plot f (from above) and

```
>  k:=t-> 10*sin(t);
```
$$k := t \rightarrow 10\sin(t)$$

execute

> `plot([f,k],-2..3, linestyle=[SOLID,DOT], thickness=[3,1]);`

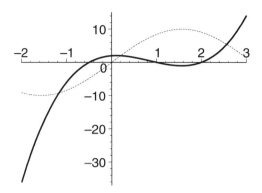

NOTE: The variable is not specified with the range.

If for some reason, you want to plot two expressions (or functions) with different ranges, then you cannot use a single `plot` command. You must make two separate plots and then combine them using the `display` command from the `plots` package. For example, to plot x^2 on $-2 \leq x \leq 1$ and $2 - x$ on $0 \leq x \leq 2$, first plot them separately and label the plots:

> `p1:=plot(x^2,x=-2..1):`

> `p2:=plot(2-x,x=0..2):`

Notice we ended these commands with colons (rather than semicolons) to suppress output. (The use of semicolons would display small uninformative outputs. Try it.) Next we load the `plots` package

> `with(plots);`

[*animate, animate3d, animatecurve, arrow, changecoords, complexplot, complexplot3d, conformal, conformal3d, contourplot, contourplot3d, coordplot, coordplot3d, densityplot, display, dualaxisplot, fieldplot, fieldplot3d, gradplot, gradplot3d, graphplot3d, implicitplot, implicitplot3d, inequal, interactive, interactiveparams, intersectplot, listcontplot, listcontplot3d, listdensityplot, listplot, listplot3d, loglogplot, logplot, matrixplot, multiple, odeplot, pareto, plotcompare, pointplot, pointplot3d, polarplot, polygonplot, polygonplot3d, polyhedra_supported, polyhedraplot, rootlocus, semilogplot, setcolors, setoptions, setoptions3d, spacecurve, sparsematrixplot, surfdata, textplot, textplot3d, tubeplot*]

NOTE: Maple responds with a list of special plot commands, one of which is `display`. (In the future, we avoid this list by using a colon instead of a semicolon. Ignore any warning.)

Now we can `display` the two plots.

> `display([p1,p2]);`

2.2. EXPRESSIONS AND FUNCTIONS: PLOTS

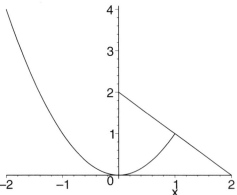

Three-Dimensional Plots: We have discussed expressions and functions of two or more variables, but the most we can plot is a function of two variables. To plot the expression

> `g:=y+sin(2*x+y);`

$$g := y + \sin(2x + y)$$

over the region $-5 \leq x \leq 5, -5 \leq y \leq 5$, use the `plot3d` command to execute

> `plot3d(g,x=-5..5,y=-5..5);`

To plot the function

> `f:=(x,y)->y+sin(2*x+y);`

$$f := (x, y) \to y + \sin(2x + y)$$

over the same region, execute

> `plot3d(f,-5..5,-5..5);`

which gives the same plot. Note the difference in syntax. For a function, you don't specify the independent variables with the intervals. However, since `f(x,y)` is an expression, it is also valid to plot `f` using

> `plot3d(f(x,y),x=-5..5,y=-5..5);`

which again gives the same plot.

As with 2-dimensional plots, there are a large number of options to modify the display of 3-dimensional plots. A complete list may be seen by executing

> `?plot3d,options`

In particular, the option `axes=normal` will add axes, `scaling=constrained` will equalize the scale on the three axes and `style=patchnogrid` will remove the grid lines.

> `plot3d(f,-5..5,-5..5, axes=normal, scaling=constrained,`
> `style=patchnogrid);`

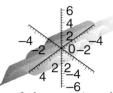

You can also modify many of these options by right-clicking in the plot and changing the option.

Several options cannot be changed from the right-click menu. For example, `grid=[#,#]` will modify the number of points computed in each direction (with the default being 25×25) and `orientation=[`θ,ϕ`]` will rotate the plot where θ and ϕ are spherical coordinates with ϕ measured down from the north pole:

```
> plot3d(f,-5..5,-5..5, grid=[5,10],
> orientation=[45,30]);
```

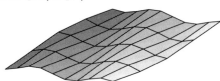

You can also *dynamically rotate the plot by clicking in the plot and dragging the mouse.* And you can see your current orientation by clicking in the plot and looking at the left of the plot toolbar. If you right click in the plot and select MANIPULATOR, you can change the action of dragging the mouse from ROTATE to SCALE or PAN. Try it!

2.3 Equations in One Variable

In Section 2.1, we defined the area of a circle as the expression

```
> Area:=Pi*r^2;
```

$$Area := \pi r^2$$

So the equation that the area of the circle is 40 may be entered and labeled by

```
> Area_eq:=Area=40;
```

$$Area\_eq := \pi r^2 = 40$$

Notice that the equality is denoted by an equal sign (=), while the assignment to a label is a colon-equal (:=). You can read off the left and right hand sides of an equation by using the `lhs` and `rhs` commands:

```
> lhs(Area_eq); rhs(Area_eq);
```

$$\pi r^2$$
$$40$$

2.3. EQUATIONS IN ONE VARIABLE

When substituting values into an equation, Maple handles the equation like an expression by using the `subs` or `eval` command. Thus if $r = a + b$, then the area equation becomes:

> `eval(Area_eq, r=a+b);`

$$\pi (a+b)^2 = 40$$

Maple has two commands for solving equations, `solve` and `fsolve`, providing exact and approximate solutions, respectively. To solve for the radius of the circle whose area is 40, execute

> `r0:=solve(Area_eq,r);`

$$r0 := \frac{2\sqrt{10}}{\sqrt{\pi}}, -\frac{2\sqrt{10}}{\sqrt{\pi}}$$

Maple does not know that r is supposed to be positive; so it has given two solutions. These may be referred to as `r0[1]` and `r0[2]`. Thus the correct solution is

> `r1:=r0[1];`

$$r1 := \frac{2\sqrt{10}}{\sqrt{\pi}}$$

To get a (floating point) decimal value, you can use `evalf`:

> `evalf(r1);`

$$3.568248232$$

Alternatively, you can get the (floating point) decimal solutions directly using

> `r2:=fsolve(Area_eq,r);`

$$r2 := -3.568248232, 3.568248232$$

and select the positive one using

> `r3:=r2[2];`

$$r3 := 3.568248232$$

Now consider a more substantial example:

EXAMPLE: Find the point(s) of intersection of the parabola $y = x^2$ and the line $y = 1 - 2x$.

SOLUTION: First let's define and plot the two expressions to get an idea where they intersect:

> `f:=x^2; g:=1-2*x;`

$$f := x^2$$
$$g := 1 - 2x$$

> `plot({f,g},x=-3..3);`

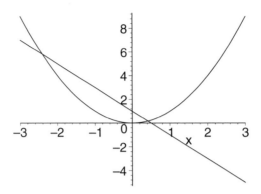

There are two solutions. To find them exactly you need to solve the equation

> f=g;
$$x^2 = 1 - 2x$$

> x0:=solve(f=g,x);
$$x0 := -1 + \sqrt{2},\ -1 - \sqrt{2}$$

Notice that we labeled the solutions x0. So the first and second solutions may be referred to as x0[1] and x0[2]. Alternatively, you can immediately give two names to the two solutions by using Maple's multiple assignment command:

> x1,x2:=solve(f=g,x);
$$x1,\ x2 := -1 + \sqrt{2},\ -1 - \sqrt{2}$$

TIP: When the result of a Maple command is two or more expressions separated by commas (called an expression sequence), you may assign them to the same number of labels separated by commas, as above.

Returning to the example, the y coordinates are

> y1:=subs(x=x1,g); y2:=subs(x=x2,g);
$$y1 := 3 - 2\sqrt{2}$$
$$y2 := 3 + 2\sqrt{2}$$

In decimals, the two points are

> pt1:=evalf([x1,y1]); pt2:=evalf([x2,y2]);
$$pt1 := [0.414213562,\ 0.171572876]$$
$$pt2 := [-2.414213562,\ 5.828427124]$$

which agrees with the plot.

Exact solutions to complicated equations can be difficult or impossible to find. For example, enter the equation

> eq:=x^3-2*x^2+x-3=0;
$$eq := x^3 - 2x^2 + x - 3 = 0$$

2.3. EQUATIONS IN ONE VARIABLE

Solving this equation with the command `solve(eq,x);` will yield three very complicated solutions. (Try this. The letter I is used to denote $\sqrt{-1}$. So two solutions are actually complex numbers.)

To get decimal approximations to these solutions, change the coefficients of the equation from exact integers to floating-point decimals.

> `eq:=x^3-2.0*x^2+x-3.0=0;`
$$eq := x^3 - 2.0\,x^2 + x - 3.0 = 0$$

Now solving the equation will yield nicer (but only approximate) decimal answers.

> `solve(eq,x);`

$2.174559410,\ -0.08727970515 + 1.171312111\,I,\ -0.08727970515 - 1.171312111\,I$

Note that the answers involving $I = \sqrt{-1}$ are complex.

Some equations are impossible to solve exactly. In fact, there is a famous theorem in mathematics that states that there is no formula (analogous to the quadratic formula) for finding roots of polynomials of fifth degree or higher. Consider the following equation.

> `eq:=x^7+3*x^4+2*x-1=0;`
$$eq := x^7 + 3\,x^4 + 2\,x - 1 = 0$$

Try finding the exact solutions using Maple.

> `sol:=solve(eq,x);`

$sol := \text{RootOf}(\%1,\ index = 1),\ \text{RootOf}(\%1,\ index = 2),\ \text{RootOf}(\%1,\ index = 3),$
$\text{RootOf}(\%1,\ index = 4),\ \text{RootOf}(\%1,\ index = 5),\ \text{RootOf}(\%1,\ index = 6),$
$\text{RootOf}(\%1,\ index = 7)$
$\%1 := \_Z^7 + 3\,\_Z^4 + 2\,\_Z - 1$

Maple's response indicates that it does not know how to solve this equation exactly. Maple knows there are seven roots and uses the notation `RootOf(...,index= #)` to denote the seven roots. In this situation, use Maple's `evalf` or `fsolve` commands to find approximate solutions.

> `evalf(sol);`

$0.4414177090,\ 0.7266822765 + 1.126611893\,I,\ 0.2452048032 + 0.8980515885\,I,$
$-1.192595934 + 0.1793080390\,I,\ -1.192595934 - 0.1793080390\,I,$
$0.2452048032 - 0.8980515885\,I,\ 0.7266822765 - 1.126611893\,I$

> `fsolve(eq,x);`

$$0.4414177090$$

Notice that `solve` with `evalf` gives all the complex roots while `fsolve` only gives the real root, unless you add the `complex` parameter to the command:

> `fsolve(eq,x,complex);`

$-1.192595934 - 0.1793080390\,I$, $-1.192595934 + 0.1793080390\,I$,
$0.2452048032 - 0.8980515885\,I$, $0.2452048032 + 0.8980515885\,I$,
0.4414177090, $0.7266822765 - 1.126611893\,I$, $0.7266822765 + 1.126611893\,I$

When the equation is not polynomial, it is best to use the `fsolve` command in conjuction with a plot. Consider the equation

> `eq:=x^2+1/x-1/x^2=0;`

$$eq := x^2 + \frac{1}{x} - \frac{1}{x^2} = 0$$

Solving this equation with `fsolve` will yield one solution.

> `sol1:=fsolve(eq,x);`

$$sol1 := -1.220744085$$

However, a plot of the expression $f := x^2 + 1/x + 1/x^2$ indicates that there is another solution.

> `plot(x^2+1/x-1/x^2,x=-2..2,y=-20..20);`

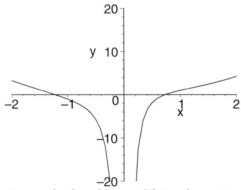

This second solution can be found by modifying the `fsolve` command to include a range:

> `sol2:=fsolve(eq,x=0..1);`

$$sol2 := 0.7244919590$$

TIP: The `fsolve` command will find all the real roots of a polynomial, but as the previous example shows, more complicated equations should be handled by a plot together with specifying a range in the `fsolve` command. Alternatively, the `avoid` option can be used to avoid known solutions:

> `sol2:=fsolve(eq,x,avoid={x=sol1});`

$$sol2 := 0.7244919590$$

2.4 Equations in Two or More Variable

Equations can involve any number of variables. For example, the equation of a circle of radius 2, centered at the origin is

> `circle:=x^2+y^2=4;`

2.4. EQUATIONS IN TWO OR MORE VARIABLE

$$circle := x^2 + y^2 = 4$$

The simple `plot` command cannot plot equations. To plot an equation involving two variables, use the `implicitplot` command in the `plots` package. (Unlike the `plot` command, the `implicitplot` command does not have a default range. The x- and y-ranges must be entered each time.)

```
> with(plots):
> implicitplot(circle, x=-2..2, y=-2..2);
```

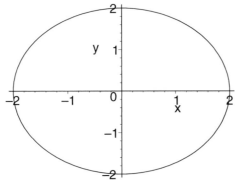

The graph of $x^2 + y^2 = 4$ should be a circle. If you get an ellipse, as above, it is because the scale is not the same on both axes. To fix this, add the option `scaling=constrained`. You can also increase the resolution by adding the option `grid=[p,q]` where `p` specifies the number of grid points along the x-axis and `q` specifies the number of grid points along the y-axis. The default values are `p=q=25`. (Don't make `p` and `q` too large so the computer does not get overloaded. `p=q=75` is usually sufficient.)

```
> implicitplot(circle, x=-2..2, y=-2..2, grid=[49,49],
> scaling=constrained);
```

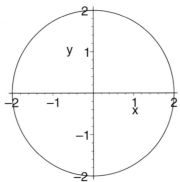

You can add the graph of a line to this plot by using curly braces { }

```
> line:=2*y=x-2;
```

$$line := 2y = x - 2$$

```
> implicitplot({circle,line}, x=-2..2, y=-2..2,
> scaling=constrained);
```

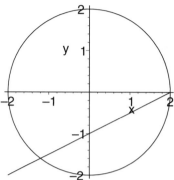

Maple can find the points of intersection. To find them approximately, click in the plot on an intersection point. The coordinates will appear in a box at the top left of the Maple window, just under the FILE menu. To find them exactly use the solve command with curly braces { }:

> sol:=solve({circle,line},{x,y});

$$sol := \{x = 2, y = 0\}, \{x = \frac{-6}{5}, y = \frac{-8}{5}\}$$

NOTE: In the solve command, Maple requires curly braces enclosing the equations {circle,line} and the variables {x,y}. The output sol consists of two solutions and each is also in the form of a set. Within each set, the variables may appear in a random order. You may refer to the two solutions as sol[1] and sol[2] but to get them in a useful form as points, execute the following:

> p1:=subs(sol[1],[x,y]); p2:=subs(sol[2],[x,y]);

$$p1 := [2, 0]$$
$$p2 := [\frac{-6}{5}, \frac{-8}{5}]$$

Maple can also solve systems with larger numbers of equations.

EXAMPLE: Solve the system of equations

$$4x - 2y + z = 8$$
$$x - 3y + 2z = 1$$
$$3x + y - 2z = 9$$

SOLUTION: Enter the equations:

> eq1:=4*x-2*y+z=8; eq2:=x-3*y+2*z=1; eq3:=3*x+y-2*z=9;

$$eq1 := 4x - 2y + z = 8$$
$$eq2 := x - 3y + 2z = 1$$
$$eq3 := 3x + y - 2z = 9$$

Solve the equations:

> sol:=solve({eq1,eq2,eq3},{x,y,z});

$$sol := \{x = 2, y = -1, z = -2\}$$

Convert the solution into a point:
> `pt:=subs(sol,[x,y,z]);`
$$pt := [2, -1, -2]$$
You can also plot equations with three variables using the `implicitplot3d` command in the `plots` package. For example, here is the plot of a sphere.
> `sphere:=x^2+y^2+z^2=9;`
$$sphere := x^2 + y^2 + z^2 = 9$$
> `implicitplot3d(sphere, x=-3..3, y=-3..3, z=-3..3,`
> `scaling=constrained);`

2.5 Summary

- Distinguish between Maple expressions, functions, and equations.

- Know how to form a Maple expression, how to use arrow notation to form a Maple function and how to enter an equation.

- Know the proper syntax to evaluate a Maple expression or equation (using `subs` or `eval`) or a Maple function (by applying it to a value).

- Convert a Maple function to an expression by applying the function to its variable. Convert a Maple expression to a function by using the `unapply` command.

- Distinguish between the assignment command which changes the values in memory and `subs` or `eval` which only affect what is on the screen. Distinguish between an assignment (:=) and an equation (=).

- Know the syntax for plotting a Maple expression using `plot` or `plot3d`, a Maple function using `plot` or `plot3d` and a Maple equation using `implicitplot` or `implicitplot3d`.

- Know how to use `solve` to find exact solutions to one or more equations and how to use `evalf` to convert exact solutions to (floating point) decimal approximations.

- When `solve` produces expressions involving `RootOf`'s, know how to use `evalf` to obtain decimal approximations.

- Know how to use `fsolve` to produce approximate (floating point) decimal solutions. Know that `fsolve` will find all real solutions for polynomial equations but only one real solution for non-polynomial equations.

- Know how to find additional decimal solutions by using a plot in conjunction with `fsolve` and a variable range or an `avoid` option.

- Be able to display two graphs simultaneously in the same plot either by enclosing several expressions or functions in curly braces or square brackets or by using the `display` command.

- Right-click in the plot window or use `plot` command options to adjust the display of plots.

2.6 Exercises

1. Enter an expression that describes the area of a square in terms of its diagonal length. Use the variable name `areasq` for the area of the square and let d be the diagonal length. Use the `subs` command to compute the area of the square when $d = 3$, $d = 3.2$ and $d = 4\pi$.

2. Enter an expression that describes the area of a circle in terms of its circumference. Use the variable name `areacir` for the area of the circle and let C be the circumference. Use the `eval` command to compute the area of the circle when $C = 3$, $C = 3.2$ and $C = 4\pi$. Find a decimal approximation using `evalf`.

3. Enter an expression that describes the volume of a cube in terms of its diagonal length. Use the variable name `vol` for the volume of the cube and let d be the diagonal length. (If you assigned a value to d in Exercise 1, be careful to unassign it before you enter this expression.) Use the `subs` command to compute the volume of the cube when $d = 3$, $d = 3.2$ and $d = 4\pi$.

4. Use `unapply` to convert the expressions from Exercises 1–3 into Maple functions. Evaluate these functions when the independent variable (either d or C) is 3, 3.2 and 4π.

    TIP: Execute `restart` now and before most exercises.

5. The volume of a cone of radius r and height h is $V = \dfrac{1}{3}\pi r^2 h$. Write the volume as an arrow defined function of the two variables r and h. Evaluate this function at $r = 4$ and $h = 3$. Convert the volume function into an expression in terms of radius R and height H.

6. Define an expression for the volume of a prism whose height is h and whose base is an equilateral triangle of side length s. Use `unapply` to convert this into a function of the two variables s and h. Compute the volume of the prism when $s = 4$, $h = 3$ and when $s = 3.1$, $h = 3.9$. Note the area of an equilateral triangle of side s is $\dfrac{\sqrt{3}s^2}{4}$.

2.6. EXERCISES

7. A large tank initially contains 200 gallons of salt water with a concentration of 1/10 lb of salt per gallon. Salt water with a concentration of 1/4 lb per gallon flows into the tank at the rate of 5 gallons per minute.

   NOTE: Concentration = Salt / Volume

   (a) Express the total amount of salt after t minutes as an expression in t; label this amount `Salt`. Use the `subs` command to evaluate `Salt` at $t = 5$, 12 and 60 minutes.

   (b) Write the volume as an expression in t and evaluate it at $t = 5$, 12 and 60 minutes.

   (c) Write the concentration as an expression in t and evaluate it at $t = 5$, 12 and 60 minutes.

8. Enter $f = \dfrac{x^5 + 1}{x^4 - 1}$ as an expression. Plot the expression f over the interval $-2 \leq x \leq 0$. Try using `subs` or `eval` to find $f(-1)$. Explain. Click in the plot at the point on the graph where $x = -1$ to find $\lim\limits_{x \to -1} f$. (When you click in the plot, the coordinates are shown at the top-left of the Maple window.) Then evaluate the limit. See ?Limit.

9. Enter $g(x) = \dfrac{x^2 - 9}{x^4 - 81}$. Plot the graph of g over the interval $1 \leq x \leq 5$. Try to find $g(3)$. Explain. Then evaluate the limit $\lim\limits_{x \to 3} g(x)$.

10. Consider the *expression* entered in Maple as `f:=x^2-1;`. Which of the following statements involve incorrect Maple syntax? How would you fix it?

    (a) `plot(f,x=-3..3);`   (b) `plot(f,-3..3);`
    (c) `plot(f(x),x=-3..3);` (This is incorrect syntax, but works anyway.)
    (d) `factor(f);`   (e) `subs(x=3,f);`   (f) `subs(x=3,f(x));`
    (g) `eval(f(x), x=3);`   (h) `f(3);`   (i) `solve(f=3,x);`
    (j) `solve(f(x)=3,x);`

11. Repeat Exercise 10 if f is now entered as the *function* `f:=x->x^2-1;`.

12. Enter each of the following as expressions: $f = \dfrac{x-3}{3x-1}$ and $g = \dfrac{x}{x+3}$. Convert each expression to a function using `unapply`. Find the composition of the two functions by entering `f(g(x))` and by entering `(f@g)(x)` and simplify each. Here, the @ symbol is used to indicate the composition, just as ∘ is used in standard mathematical notation. Note that the output is an expression, since each function has been evaluated. Also try entering `h:=f@g;` which produces a function and then evaluating at x.

13. Graph $g = \dfrac{\sqrt{4-x} - \sqrt{3+x}}{x^2 - 4}$ showing only $-5 \leq x \leq 5$ and $-10 \leq y \leq 10$. Notice the vertical line segments that Maple puts in at the vertical asymptotes. Right click in the plot and select STYLE > POINT to see only the points Maple used and so verify that the lines are not really there. Now add the option discont=true to the plot, both without and with the option style=point. What happened?

14. Enter the expression $g = \dfrac{\sqrt{4-x} - \sqrt{3+x}}{x^2 - 4}$.

    (a) Change the expression g into a function f by using unapply.
    (b) Find $f(1), f(2), f(5)$.
    (c) Find $g(1), g(2), g(5)$ using subs.
    (d) Discuss what happens at $x = 2$ and $x = 5$.

15. Form the expression $f = \sqrt{x^2 \sin(x + \pi/4)}$.

    (a) Plot this expression for $-2 \leq x \leq 4$ and $-1 \leq y \leq 2$. Click on the endpoints of the pieces to determine the domain of the expression. (When you click in the plot, the coordinates are shown in the top-left of the Maple window.) If the axes get in the way, right click in the plot and select AXES > FRAMED to move them.
    (b) Determine the domain of the expression to ten decimal places by using fsolve on $f = 0$, with ranges for x found from the plot.

16. Solve the following equations using the solve command. In each case, check your answers with a plot and by substituting the roots into the equation.

    (a) $x^2 - 4x + 1 = 0$
    (b) $x^2 - 4.0x + 1.0 = 0$
    (c) $x^2 - x + 2 = 0$
    (d) $x^2 - x + 2.0 = 0$
    (e) $x^3 + 2x + 3 = 0$
    (f) $x^3 + 2.0x + 3.0 = 0$

17. Find all solutions to the following equations over the given interval.

    (a) $\sin^2(x) = \cos(4x)$, for $0 \leq x \leq \pi$
    (b) $\cos^2(x) = \sin(4x)$, for $0 \leq x \leq \pi$
    (c) $8\cos(x) = x$, for $-\infty < x < \infty$

    HINT: Use the plot command to graph the two sides of each equation and click on each intersection. Then get a more accurate answer by using fsolve.

2.6. EXERCISES

18. Solve the system of equations $3x - 2y = 4$ and $2x + 5y = 4$ for x and y.

19. Find the intersection point of the lines $y = -\frac{3}{4}x + \frac{1}{4}$ and $y = -\frac{1}{4}x + \frac{3}{4}$.

20. Find the decimal approximations for all real roots of the equation

$$\frac{1}{x^4} + x^2 - 2x - 6 = 0$$

(Make sure you plot this one).

21. For each pair of expressions, plot them separately over the given intervals with different colors. Then combine them into a single plot using `display`.

    (a) $f := -x^4$ for $-1 \leq x \leq 1$ and $g := 3x - 2$ for $0 \leq x \leq 2$
    (b) $f := x^3 - 2x^2$ for $-2 \leq x \leq 3$, and $g := x^4 - 3x$ for $-1 \leq x \leq 3$
    (c) $f := \sin(x) + \cos(x)$ for $-2\pi \leq x \leq \pi$ and $g := \sin(x) - \cos(x)$ for $-\pi \leq x \leq 2\pi$

22. Define the expression $f = 25x + \dfrac{1}{x - 25}$ in Maple. Enter the following plot commands and discuss the behavior of each plot near $x = 25$.

    (a) `plot(f,x=0..100);`
    (b) `plot(f,x=0..100, numpoints=100);`
    (c) `plot(f,x=0..100, numpoints=10000);`

23. Use the `plot3d` command to view the graphs of the following expressions. Then click in the plot and drag the mouse.

    (a) $f = x^2 + y^2$ (b) $f = y^2 - x^2$ (c) $f = x^4 - y^4$
    (d) $f = (x^2 + y^2)(x - y)$ (e) $f = \cos(x - y)$ (f) $f = \sin(x)\sin(y)$

24. The point of this exercise is that plots can be deceiving. Enter the expression $f := x^3 - x^2 - x + 1.001$. Plot f over the interval $-2 \leq x \leq 2$. From this plot, guess how many real solutions there are to the equation $f = 0$. Now solve the equation $f = 0$ with the `solve` command. How many solutions did `solve` return? Now replot f over the interval $0.9 \leq x \leq 1.1$. Does the graph of f cross the x-axis near the point $x = 1$?

25. The point of this exercise is to show how piecewise defined functions can be entered into Maple using the `piecewise` command. The function

$$f(x) = \begin{cases} x^2 + 1 & x < 1 \\ 3 & x = 1 \\ x - 1 & \text{otherwise} \end{cases}$$

can be entered into Maple as either the expression:

```
> f:=piecewise(x<1, x^2+1, x=1, 3, x-1);
```

or as the function

> `g:=x->piecewise(x<1, x^2+1, x=1, 3, x-1);`

The arguments of `piecewise` come in pairs consisting of a condition and a value, except that the last (optional) argument is the "otherwise" value. The expression `f` and the function `g` can now be evaluated. For example, the value of `f` at $x = .5$ is

> `eval(f,x=.5);`

and the value of `g` at $x = 1$ is

> `g(1);`

To plot `f` and `g` over the interval $-1 \leq x \leq 3$ enter

> `plot(f, x=-1..3, discont=true);`
> `plot(g, -1..3, discont=true);`

NOTE: Maple is not reliable about plotting single points: There is a circle at $(1, 3)$ in the graph of `f` but not in the graph of `g`.

Enter and plot the following functions with the `axes=framed` option.

$$p(x) = \begin{cases} \sin(x) & \text{if } x < 0 \\ \frac{1}{2} & \text{if } x = 0 \\ \cos(x) & \text{if } x > 0 \end{cases}$$

$$q(x) = \begin{cases} x^2 & \text{if } x \leq 0 \\ \sqrt{x} & \text{if } x > 2 \\ -2x & \text{otherwise} \end{cases}$$

NOTE: $\leq$ is entered as `<=`.

26. Enter into Maple the equation of the circle with center at $x = 2$ and $y = 3$ with radius 2. Then graph the circle in two ways:

    (a) `solve` the equation for y and label the solution. Plot the two solutions simultaneously using a single `plot` command. (Note that the circle looks more like an ellipse than a circle. What can you do to remove the distortion?)

    (b) Use `implicitplot` in the `plots` package to plot the equation directly.

Chapter 3

Data Sets and Parametric Curves

The chapter discusses ways to describe mathematical quantities other than expressions, functions and equations. It begins with the Maple commands to plot data points and fit a polynomial curve to a given set of data both exactly and approximately. Then it describes how to plot parametric curves, with applications to inverse functions and polar curves.

This chapter can be safely skipped until data sets, parametric curves, inverse functions or polar curves are needed.

3.1 Data Points and Curve Fitting

Often, problems involve data points rather than functions or expressions or equations. You may want to plot these points, and you may want to find a curve which passes through these points, either exactly or approximately.

Data Sets: In Maple, an ordered list must be enclosed in square brackets [ ]. So [1,2.2] is an ordered pair, or point, while [[1,2.2], [3,5.5], [6,4.2]] is an ordered list of points. To plot this list of points, first assign it to a variable, say *mydata*

```
>   mydata:=[[1,2.2], [3,5.5], [6,4.2]];
```
$$mydata := [[1, 2.2], [3, 5.5], [6, 4.2]]$$
and check for typing errors. Then issue the `plot` command

```
>   plot(mydata, style=point);
```

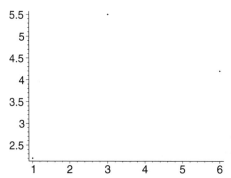

NOTE: You have already seen point plots using the `style=point` option in Section 1.4 when we discussed plots of discontinuous functions.

You can change the x- and y-ranges for the plot, either explicitly (as the second and third options) or by including a `view` option. You can change the symbol used for the points and the size of the symbol by using the `symbol` and `symbolsize` options. For example:

```
> plot(mydata,0..7,0..6, style=point, symbol=cross,
> symbolsize=24);
```

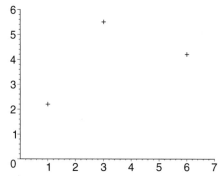

Deleting the option `style=point` will draw this data set with connecting line segments.

```
> plot(mydata, view=[0..7,0..6]);
```

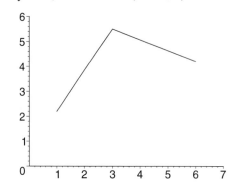

3.1. DATA POINTS AND CURVE FITTING

Exact Curve Fitting: Looking at the plot, or for other physical reasons, you might want to connect the dots with a parabola instead of line segments.

EXAMPLE 1: Find the expression whose graph is the parabola that passes through the points $(1, 2.2)$, $(3, 5.5)$, and $(6, 4.2)$. Then graph the parabola and the points in the same plot.

SOLUTION: First, enter the general formula for a parabola (as a function of x).

```
>   p:=x->a*x^2+b*x+c;
```
$$p := x \rightarrow a\,x^2 + b\,x + c$$

The unknown coefficients a, b, and c must be found so that the parabola passes through the points $(1, 2.2)$, $(3, 5.5)$, and $(6, 4.2)$. In order for the parabola to pass through the point $(1, 2.2)$, the equation $p(1) = 2.2$ must be satisfied. We enter this equation and label it *eq1*:

```
>   eq1:=p(1)=2.2;
```
$$eq1 := a + b + c = 2.2$$

Similarly, the other two equations are.

```
>   eq2:=p(3)=5.5; eq3:=p(6)=4.2;
```
$$eq2 := 9\,a + 3\,b + c = 5.5$$
$$eq3 := 36\,a + 6\,b + c = 4.2$$

These equations are then solved and the solution is labeled *sol*.

```
>   sol:=solve({eq1,eq2,eq3},{a,b,c});
```
$$sol := \{a = -0.4166666667,\ b = 3.316666667,\ c = -0.7000000000\}$$

(The order of the variables in your solution may differ.) These values of a, b, and c can be substituted into $p(x)$ using the **subs** command.

```
>   q:=subs(sol,p(x));
```
$$q := -0.4166666667\,x^2 + 3.316666667\,x - 0.7000000000$$

Here the label q is given to the parabola. Note that **q** is defined as a *Maple expression*.

NOTE: As mentioned in Sections 1.4 and 2.2, two or more expressions (or functions) can be plotted on the same interval by enclosing the expressions (or functions) in curly braces { }. However, if the variables or intervals are different, or if they are two different kinds of plots (expressions vs. functions vs. equations vs. points) then you must use the **display** command from the **plots** package.

To plot the parabola and the points, we first plot them separately and label the plots:

```
>   p1:=plot(mydata,0..7,0..6, style=point, symbol=cross,
>   symbolsize=36):
```

```
>   p2:=plot(q,x=0..7):
```

Notice we ended these commands with colons (rather than semicolons) to suppress output. Next we load the plots package and display the two plots (this time with a semicolon).

> with(plots): display([p1,p2]);

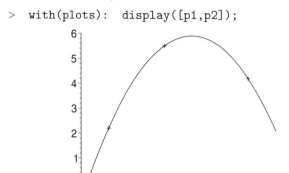

The points are right on the parabola.

Approximate Curve Fitting: If there are many more points, you may want to find a curve which best fits the data but does not necessarily go precisely through the points. This is simple using Maple's stats and statplots packages.

EXAMPLE 2: Consider the data points given in the table below:

x	1	2	3	4	5	6	7
y	2.2	3.7	5.5	5.9	5.6	4.2	3.1

Find a parabola which best fits the data. Then graph the parabola and the points in the same plot.

SOLUTION: Load the stats and statplots packages and define lists of the x- and y-coordinates in the data.

> with(stats): with(statplots):

> xs:=[1,2,3,4,5,6,7]; ys:=[2.2,3.7,5.5,5.9,5.6,4.2,3.1];

$$xs := [1, 2, 3, 4, 5, 6, 7]$$
$$ys := [2.2, 3.7, 5.5, 5.9, 5.6, 4.2, 3.1]$$

Plot the data using the scatterplot command from the statplots package.

> p1:=scatterplot(xs,ys, symbol=cross, symbolsize=24): p1;

3.1. DATA POINTS AND CURVE FITTING

Although this method of plotting points might seem more complicated, it has some advantages. For example, Maple's built-in curve fitting commands require the x- and y-coordinates to be entered separately.

From the plot, it appears as though a quadratic function might give a good approximation to this data. The fit[leastsquare[...]] command from the stats package will create a quadratic least squares fit.

> fit[leastsquare[[x,y],y=a*x^2+b*x+c]]([xs,ys]);

$$y = -0.3619047619\,x^2 + 3.030952381\,x - 0.5714285714$$

> q:=rhs(%);

$$q := -0.3619047619\,x^2 + 3.030952381\,x - 0.5714285714$$

Let's see how well we did.

> p2:=plot(q,x=0..8):
> display({p1,p2});

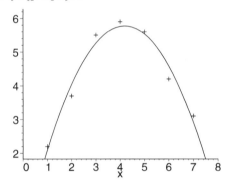

Not bad.

The fit[leastsquare[...]] command can be easily modified to fit other types of curves.

EXAMPLE 3: Consider the population data given in the table below.

Date	1900	1910	1920	1930	1940	1950
Pop. (billions)	1.65	1.75	1.86	2.07	2.30	2.52

1960	1970	1980	1990	1996
3.02	3.70	4.45	5.30	5.77

Enter this data into Maple, obtain a cubic fit and plot the data with the cubic function.

SOLUTION: The time values are in decades since 1900.

> ts:=[0,1,2,3,4,5,6,7,8,9,9.6]:
> Ps:=[1.65,1.75,1.86,2.07,2.30,2.52,3.02,3.70,4.45,5.30,5.77]:
> fit[leastsquare[[t,P],P=a*t^3+b*t^2+c*t+d]]([ts,Ps]);

$$P = 0.002325670340\,t^3 + 0.01914440256\,t^2 + 0.03583942511\,t + 1.676870710$$

> f:=rhs(%);

$$f := 0.002325670340\,t^3 + 0.01914440256\,t^2 + 0.03583942511\,t + 1.676870710$$

```
> p1:=scatterplot(ts, Ps, symbol=cross, symbolsize=24):
> p2:=plot(f,t=0..10):
> display({p1,p2});
```

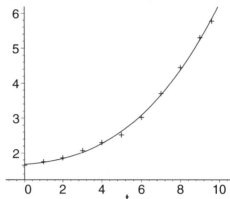

Again a good fit.

3.2 Parametric Curves

Some curves, such as circles and ellipses, are not the graphs of functions. Other curves are not even conveniently given by equations. Instead, these figures are more conveniently described by *parametric equations*, which are of the form

$$x = f(t), \quad y = g(t), \quad \text{for} \quad a \leq t \leq b$$

where f and g are functions of the parameter t, and a and b are the initial and final values of t. If you want, you can think of the equation $(x,y) = (f(t), g(t))$ as giving the position (x,y) of a particle in a plane as a function of time t. However, in general, t may not represent time (and you do not need to use the letter t for the parameter). It may represent some other physical or geometrical quantity such as angle or arclength.

To plot the parametric curve $x = f(t)$, $y = g(t)$, define the component functions as Maple expressions f and g and type the command plot([f,g,t=a..b]). NOTE: It is easy to confuse the syntax for a parametric plot with the square bracket syntax for plotting two expressions at the same time. For parametric plots, square brackets [ ] must be used and the parameter and its range must be *inside* the brackets. To plot two functions, you may use square brackets [ ] but the variable and range must be *outside* the brackets.

EXAMPLE 1: Plot the circle of radius 3 centered at $[2, 1]$.

SOLUTION: The circle may be parametrized by:

$$x = 2 + 3\cos(\theta) \quad y = 1 + 3\sin(\theta), \quad \text{for} \quad 0 \leq \theta \leq 2\pi$$

where the parameter is the angle θ. To plot the circle we enter the coordinates as Maple expressions and plot it:

3.2. PARAMETRIC CURVES

```
> f,g:=2+3*cos(theta),1+3*sin(theta);
```
$$f, g := 2 + 3\cos(\theta), 1 + 3\sin(\theta)$$

```
> plot([f,g,theta=0..2*Pi], scaling=constrained);
```

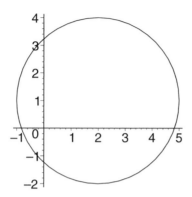

NOTE: We have included the scaling=constrained option to equalize the scales on the axes.

EXAMPLE 2: *Lissajous figures* A Lissajous figure is a curve in the plane which may be parametrized as

$$(x, y) = (\cos(pt), \sin(qt))$$

where p and q are positive integers which are relatively prime (no common factors other than 1). Plot the Lissajous figure with $p = 5$ and $q = 4$.

SOLUTION: Here it is:

```
> plot([cos(5*t),sin(4*t),t=0..2*Pi]);
```

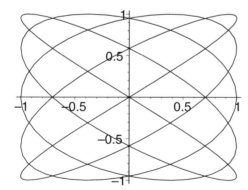

As t varies from 0 to 2π, the x-coordinate oscillates $p = 5$ times while the y-coordinate oscillates $q = 4$ times. (There are 5 bumps on the left and right and 4 bumps on the top and bottom.)

3.3 Inverse Functions

The inverse of a function f is a function f^{-1} which undoes f. Thus

$$y = f^{-1}(x) \quad \text{means} \quad x = f(y)$$

For example, the cube root is the inverse of the cube function.

EXAMPLE 1: Find the inverse of the function $f(x) = \sqrt[5]{(x+4)^3 - 7}$. Then graph the function, the inverse and the 45° line $y = x$ in the same plot.

SOLUTION: Define the equation $y = f(x)$:

> `eq1:=y=((x+4)^3-7)^(1/5);`

$$eq1 := y = ((x+4)^3 - 7)^{(1/5)}$$

Interchange x and y: (The braces make the substitutions occur simultaneously.)

> `eq2:=subs({x=y,y=x},eq1);`

$$eq2 := x = ((y+4)^3 - 7)^{(1/5)}$$

Solve for y:

> `sol:=solve(eq2,y);`

$$sol := (7+x^5)^{(1/3)} - 4, \; -\frac{(7+x^5)^{(1/3)}}{2} + \frac{1}{2} I \sqrt{3} \,(7+x^5)^{(1/3)} - 4,$$
$$-\frac{(7+x^5)^{(1/3)}}{2} - \frac{1}{2} I \sqrt{3} \,(7+x^5)^{(1/3)} - 4$$

We only want the real solution. So

> `eq3:=y=sol[1];`

$$eq3 := y = (7+x^5)^{(1/3)} - 4$$

Now we can plot the two functions and the diagonal:

> `plot([rhs(eq1),rhs(eq3),x], x=-2..4, y=-2..4,`
> `linestyle=[SOLID,DASH,DOT], thickness=[5,3,1],`
> `scaling=constrained);`

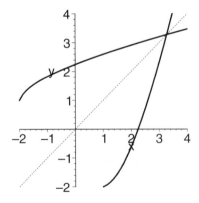

3.3. INVERSE FUNCTIONS

Notice that the dashed inverse function is the mirror image (through the diagonal) of the original solid function. This is because the inverse function is obtained by interchanging x and y (and solving for y).

Frequently, it is impossible to solve explicitly for an inverse function. However, it is still possible to plot the inverse function using a parametric plot. Given a function $y = f(x)$, its parametric form is $x = t$, $y = f(t)$. Since the inverse function $y = f^{-1}(x)$ satisfies $x = f(y)$, its parametric form is $x = f(t)$, $y = t$.

EXAMPLE 2: Plot the function $f(x) = 4 + x + \sin(x)$ and its inverse.
NOTE: It is impossible to solve the equation $x = 4 + y + \sin(y)$ for y.
SOLUTION: We enter the function, but as an expression in t:

```
>   f:=4+t+sin(t);
```
$$f := 4 + t + \sin(t)$$

Then the parametric plot of the function is

```
>   plot([t,f,t=-10..10], view=[-10..10,-10..10],
>   scaling=constrained);
```

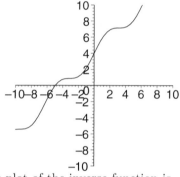

and the parametric plot of the inverse function is

```
>   plot([f,t,t=-10..10], view=[-10..10,-10..10],
>   scaling=constrained);
```

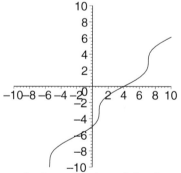

Notice that the inverse is the reflection of the function through the diagonal.

More generally, if $x = f(t)$, $y = g(t)$ is the parametric form of a function, then $x = g(t)$, $y = f(t)$ is the parametric form of its inverse function.

Sometimes texts obfuscate the issue of inverses by placing undue emphasis on whether or not the (original or) inverse is a function, i.e., passes "the vertical line test." As shown below, one can *always* graph the inverse of a "relation" given by an equation; one simply reverses the variables and uses `implicitplot`. In the following example, the original curve is a function since it passes the vertical line test, but the original is not 1-1 since it does not pass the horizontal line test. So its inverse relation (with the variables flipped) will not pass a vertical line test. So the inverse curve exists, but is not a function. We still have symmetry with respect to the 45° line and both graphs are easy to draw using `implicitplot`.

EXAMPLE 3: Plot the relation $x^2 + y^3 = 1$ and its inverse and the 45° line in one plot. Give the different relations different colors and thicknesses. Notice that the original relation is a function. Then select a maximal interval on which the original function is 1-1. What is the domain of its inverse? Plot them both with the 45° line in one plot.

SOLUTION: We enter the relation, interchange the variables to get the inverse relation and plot them:

```
> eqn:=x^2+y^3=1;
> inv:=subs({x=y,y=x},eqn);
```

$$eqn := x^2 + y^3 = 1$$
$$inv := y^2 + x^3 = 1$$

```
> with(plots):
> implicitplot([eqn,inv,y=x],x=-3..3,y=-3..3, scaling=constrained,
> thickness=[1,3,1], color=[red,blue,black]);
```

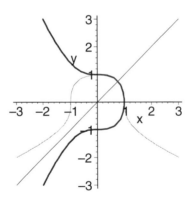

Select the interval $[0, \infty)$. The domain of the inverse will be $(-\infty, 1]$. We plot them: (We plot using a 3 instead of ∞.)

```
> p1:=implicitplot(eqn,x=0..3,y=-3..1, scaling=constrained,
> thickness=1, color=red):  p1;

> p2:=implicitplot(inv,x=-3..1,y=0..3, scaling=constrained,
> thickness=3, color=blue):  p2;

> p3:=implicitplot(y=x,x=-3..3,y=-3..3):  p3;

> display([p1,p2,p3]);
```

3.4. POLAR CURVES

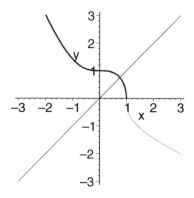

3.4 Polar Curves

Polar coordinates (r, θ) are related to rectangular coordinates (x, y) by

$$x = r\cos\theta \qquad y = r\sin\theta$$

$$r = \sqrt{x^2 + y^2} \qquad \tan\theta = \frac{y}{x}$$

A polar curve is a special case of a parametric curve in which the radius r is given as a function of the angle θ which serves as the parameter. So if $r = f(\theta)$ is a polar equation, then the parametric equations are

$$x = r\cos(\theta) = f(\theta)\cos(\theta) \qquad y = r\sin(\theta) = f(\theta)\sin(\theta).$$

There are three ways to plot a polar curve: as an ordinary parametric curve, by including the coords=polar option in a plot command, or by using the polarplot command from the plots package.

EXAMPLE 1: Plot the polar equation

$$r = \cos(3\theta) \qquad \text{for} \quad 0 \le \theta \le 2\pi$$

SOLUTION: To plot the polar equation $r = \cos(3\theta)$, convert it to parametric form as

```
> r1:=cos(3*theta);
```
$$r1 := \cos(3\,\theta)$$

```
> x1,y1:=r1*cos(theta),r1*sin(theta);
```
$$x1,\ y1 := \cos(3\,\theta)\cos(\theta),\ \cos(3\,\theta)\sin(\theta)$$

and plot it as a parametric curve:

```
> plot([x1,y1,theta=0..2*Pi], scaling=constrained);
```

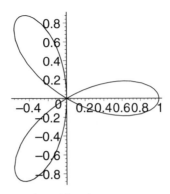

The exact same curve is produced by the plot command with the coords=polar option:

> `plot(r1,theta=0..2*Pi, coords=polar, scaling=constrained);`

and by the polarplot command from the plots package:

> `with(plots):`
> `polarplot(r1,theta=0..2*Pi, scaling=constrained);`

EXAMPLE 2: Show that the polar equation $r = \dfrac{1}{1 - \sin\theta}$ is a parabola and plot it.

SOLUTION: Enter r as an expression and as an equation.

> `r1:=1/(1-sin(theta)); eq1:=r=r1;`

$$r1 := \frac{1}{1 - \sin(\theta)}$$

$$eq1 := r = \frac{1}{1 - \sin(\theta)}$$

Substitute $\sin(\theta) = \dfrac{y}{r}$ and $r = \sqrt{x^2 + y^2}$, and solve for y: (Notice the substitutions are done successively, not simultaneously, by leaving out the braces.)

> `eq2:=subs(sin(theta)=y/r, r=sqrt(x^2+y^2), eq1);`

$$eq2 := \sqrt{x^2 + y^2} = \frac{1}{1 - \dfrac{y}{\sqrt{x^2 + y^2}}}$$

> `sol:=solve(eq2,y);`

$$sol := x\,I,\ -I\,x,\ \frac{x^2}{2} - \frac{1}{2}$$

Ignore the imaginary solutions. So the equation is

> `y=sol[3];`

$$y = \frac{x^2}{2} - \frac{1}{2}$$

which is a parabola. Now plot it for $0 \leq \theta \leq 2\pi$.

3.5. SUMMARY   49

```
>   polarplot(r1,theta=0..2*Pi);
```

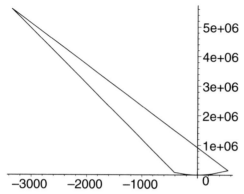

Notice that there are extraneous lines and the ranges are way too large due to the fact that the expression r is large for θ near $\pi/2$. To view a portion of the graph near the origin, add the `view` option:

```
>   polarplot(r1,theta=0..2*Pi, view=[-4..4,-3..5]);
```

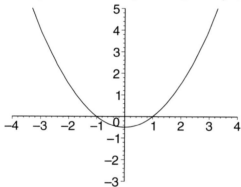

3.5 Summary

- Given a set of data as a list of points
  ```
  >   mydata:=[[ , ], [ , ], ..., [ , ]];
  ```
  plot it using
  ```
  >   plot(mydata, style=point);
  ```
  and be able to find a polynomial passing through these points.

- Given a set of data as two lists of x- and y-coordinates
  ```
  >   xs:=[ , , ..., ]; ys:=[ , , ..., ];
  ```
  plot it using
  ```
  >   with(stats):   with(statplots):
  >   p1:=scatterplot(xs,ys):  p1;
  ```
  Also find the curve which best fits the data using

```
>   fit[leastsquare[[x,y],y=expr]]([xs,ys]);
>   q:=rhs(%);
```

and plot it with the data set using

```
>   p2:=plot(q,x):
>   display({p1,p2});
```

- Given a parametric curve

$$x = f(t) \quad y = g(t), \quad \text{for} \quad a \le t \le b$$

plot it using

```
>   plot([f,g,t=a..b], scaling=constrained);
```

where f and g are expressions in t.

- The inverse of a function f is the function f^{-1} defined by

$$y = f^{-1}(x) \quad \text{means} \quad x = f(y)$$

To find f^{-1} start with the equation

```
>   eq1:=y=expr;
```

where expr is an expression in x. Interchange x and y:

```
>   eq2:=subs({x=y,y=x},eq1);
```

and solve for y:

```
>   eq3:=y=solve(eq2,y);
```

(It may be necessary to select the real solution.) Then the function and its inverse and the diagonal may be plotted using

```
>   plot([rhs(eq1),rhs(eq3),x], x=a..b, y=c..d,
>   scaling=constrained);
```

- If a parametric curve $x = f(t)$, $y = g(t)$ is the graph of a function, then the inverse function is given parametrically as $x = g(t)$, $y = f(t)$. The function and its inverse may be plotted using

```
>   plot({[f,g,t=a..b],[g,f,t=a..b]}, view=[c..d,c..d],
>   scaling=constrained);
```

- The polar curve

$$r = f(\theta) \quad \text{for} \quad a \le \theta \le b$$

may be plotted using

```
>   polarplot(f, theta=a..b, scaling=constrained);
```

where f is an expression in θ.

3.6 Exercises

1. Find the cubic polynomial whose graph passes through the points $(2, 3.2)$, $(3, 2.6)$, $(5, 1.2)$, and $(7, 3.3)$. Graph the points and your answer and put them into one plot using the `display` command.

2. *Draw a map of Texas.* To plot a map of Texas, enter the following two lists for the northern and southern boundaries of Texas.
   ```
   > north:=[[0,0], [3,0], [3,4.5], [6,4.5], [6,2.2], [7,2.1],
   >    [ 8,1.8], [9,1.9], [10,1.8], [11,1.7], [11,-2.2]];
   > south:=[[0,0], [1,-1.1], [2,-2.5], [3,-2.9], [4,-2.3],
   >    [5,-2.8], [6,-4.4], [7,-5.8], [8,-6.1], [9,-3.3],
   >    [10,-2.8], [11,-2.2]];
   ```
   Here, the origin is the westernmost corner of Texas (near El Paso) and the x-axis is the extension of the east-west border between New Mexico and Texas. Each unit represents approximately 69 miles. After entering these lists, execute:
   ```
   > plot({north, south});
   ```

3. The general equation of a circle is
   $$x^2 + y^2 + ax + by = c$$
   Find the equation of the circle (i.e., find a, b, and c) that passes through the three points $(2, 1)$, $(4, -1)$, and $(1, 7)$. Plot the circle with the points. HINT: It is not possible to graph the circle with a function of the form $y = f(x)$ (why?); so it is easier to work with expressions using the `subs` command to obtain the necessary three equations. Start by assigning the above equation to the label `circle`. Substitute the three points into the circle to obtain three equations such as `eq1:=subs(x=2,y=1,circle);`. Then solve these three equations for a, b, and c and substitute them into the circle. Finally, plot the three points using `plot` and your circle using `implicitplot` and combine them using `display`. Don't forget you need `with(plots)`.

4. Suppose a rubber band is stretched and the following data is recorded relating the restoring force to displacement.

Disp. (meters)	.02	.03	.05	.06	.09	.11	.12
Force (Newtons)	.10	.17	.28	.35	.52	.57	.65

.14	.16	.19	.22	.25	.27
.71	.77	.86	.98	1.04	1.10

   Determine whether this data is best approximated using a linear, quadratic or cubic fit. In each case, plot the data against the curve.

5. The following table lists a measurement of the specific heat of air at various temperatures.

Temperature (K)	300	400	500	600
Sp. Heat (J/gm-K)	1.0045	1.0134	1.0296	1.0507

700	800	900
1.0743	1.0984	1.1212

   Determine a linear curve that approximates this data, and plot the curve and the data in the same graph. In addition, compute the predicted values for specific heat (given by the linear approximation) corresponding to the temperature values in the table. Is there a relationship between the average of the predicted values and the average of the specific heat values in the table?

6. The following table lists a measurement of the specific heat of air at higher temperatures.

Temperature (K)	1000	1500	2000	3000
Sp. Heat (J/gm-K)	1.1910	1.2095	1.2520	1.2955

   Determine a quadratic curve that approximates this data. Plot the curve and the data in the same graph. In addition, compute the predicted values for specific heat (given by the quadratic approximation) corresponding to the temperature values in the table. Is there a relationship between the average of the predicted values and the average of the specific heat values in the table?

7. Find the exact cubic curve passing through the data points of Exercise 6, and plot the cubic curve and the data in the same graph. Does the quadratic curve of Exercise 6 or the cubic curve of this exercise give a better description of the data? Be sure to consider extrapolating to higher and lower temperatures.

8. The following Maple command makes it easy to generate random data sequences of integers between −100 and 100.

   ```
   > r:=rand(-100..100):
   ```

   For example, if you want to create random data sequences named xs and ys containing 10 data values each, simply do the following:

   ```
   > xs:=[seq(r(),i=1..10)];
   > ys:=[seq(r(),i=1..10)];
   ```

   NOTE: The values obtained will be different each time you execute these commands.

   Use Maple to create a random pair of data sequences named xs and ys containing 10 data values each. Find a linear curve that approximates this data and compute the predicted y values (given by the linear approximation) for the corresponding listed x values in xs. Compare the average

3.6. EXERCISES

of the predicted y values to the given values in ys. Repeat this process 5 times. What do you conclude?

9. The ellipse $\frac{x^2}{9} + \frac{y^2}{16} = 1$ may be parametrized by $x = 3\cos(t)$ and $y = 4\sin(t)$. Check this by substituting the parametrization into the equation. Then, plot the ellipse using a parametric plot with constrained scaling.

10. Show that the parametric curve $x = 3\cosh(t)$, $y = 4\sinh(t)$ is part of the hyperbola $\frac{x^2}{9} - \frac{y^2}{16} = 1$. Then plot the piece for $-2 \leq t \leq 2$. Recall that

$$\cosh(t) = \frac{e^t + e^{-t}}{2}, \quad \sinh(t) = \frac{e^t - e^{-t}}{2} \quad \text{and} \quad \cosh^2(t) - \sinh^2(t) = 1$$

11. For each function, find the inverse function and plot the function and its inverse in the same plot on the given interval with the inverse thicker.

   (a) $f(x) = \sqrt{144 - 9x^2}$ for $0 \leq x \leq 12$
   (b) $f(x) = \sqrt[3]{1000 - 125x^3}$ for $0 \leq x \leq 10$
   (c) $f(x) = x\,e^x$ for $-1 \leq x \leq 1$ NOTE: e^x is entered as exp(x). Also LambertW is defined as precisely this inverse function.

12. For each function, plot the function and its inverse in the same plot with the inverse thicker.

   (a) $y = x + \cos(x)$ for $-2\pi \leq x \leq 2\pi$
   (b) $y = x^3 + 16x$ for $-2 \leq x \leq 2$
   (c) $x = t^3 + 2t$, $y = t^3 + 3t$ for $0 \leq t \leq 2$

13. Graph each of the following equations by hand and then check your answers with a Maple plot. You may need to restrict the view on (a).

   (a) $r = \dfrac{4}{1 + \cos(\theta)}$
   (b) $r = 1 + 3\cos(\theta)$
   (c) $r = 4\sin(3\theta)$
   (d) $r = 3\cos(4\theta)$
   (e) $r^2 = 1 + \sin^2(\theta)$
   (f) $r = \theta$, for $\dfrac{-3\pi}{2} \leq \theta \leq \dfrac{3\pi}{2}$ (Happy Valentine's Day)

Chapter 4

Differentiation

We start this chapter by computing derivatives using the limit definition of the derivative. Then we introduce the Maple syntax for differentiation. Maple makes a distinction between expressions and functions (See Chapter 2.), and we discuss the syntax for differentiating both. We conclude this chapter with sections on implicit differentiation and linear approximation.

4.1 The Limit of the Difference Quotient

The definition of the derivative of the function $f(x)$ at the point $x = a$ is given by

$$f'(a) = \lim_{h \to 0} \frac{f(a+h) - f(a)}{h}$$

The motivation of this definition is that the derivative of f at $x = a$ should be the slope of the tangent line to the graph of f at $x = a$. The key idea is that the quantity $\dfrac{f(a+h) - f(a)}{h}$ is the slope of the line segment connecting the points $(a, f(a))$ and $(a+h, f(a+h))$. As h approaches zero, the point $(a+h, f(a+h))$ approaches $(a, f(a))$, the line segment rotates into the tangent line and the slope of the line segment approaches the slope of the tangent.

EXAMPLE: Consider the function $f(x) = x^3 - 8x$. Compute $f'(2)$ and the equation of the tangent line. Then plot the function and its tangent line in the same plot.

SOLUTION: Enter f as a Maple function and calculate the above limit with $a = 2$.

```
>   f:=x->x^3-8*x;
```
$$f := x \to x^3 - 8x$$

```
>   Limit((f(2+h)-f(2))/h,h=0); m:=value(%);
```
$$\lim_{h \to 0} \frac{(2+h)^3 - 8 - 8h}{h}$$

4.1. THE LIMIT OF THE DIFFERENCE QUOTIENT

$$m := 4$$

Note that the `Limit` command displays the limit so that you can check for typing errors and then the `value(%)` command evaluates this limit (recall that the percent % refers to the result of the previous command, which in this case is the limit).

So the slope of the tangent line at the point $(2, f(2)) = (2, -8)$ is $m = 4$. Accordingly, its formula is

> `y:=m*(x-2)+f(2);`

$$y := 4x - 16$$

This tangent line and the function can be plotted with two colors using

> `plot([y,f(x)],x=0..4, color=[green,blue]);`

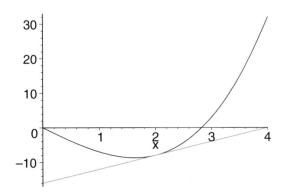

To see that this tangent line closely approximates the graph of the function f near $x = 2$, replot this graph over a small interval about $x = 2$.

> `plot([y,f(x)],x=1.9..2.1, color=[green,blue]);`

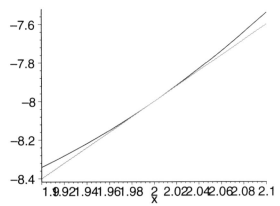

Another way to obtain $f'(2)$ is to compute $f'(x)$ and then substitute $x = 2$. To do this, compute the above limit with $a = x$ and store the result in the variable named `Df`.

```
>   Limit((f(x+h)-f(x))/h,h=0); Df:=value(%);
```
$$\lim_{h \to 0} \frac{(x+h)^3 - 8h - x^3}{h}$$
$$Df := 3x^2 - 8$$

To obtain $f'(2)$, evaluate Df at $x = 2$:
```
>   eval(Df, x=2);
```
$$4$$

which is the same slope as above.

Let's examine the above limit process more carefully. First, enter the difference quotient $\dfrac{f(x+h) - f(x)}{h}$ into Maple. Store this expression as the variable diffq and simplify.
```
>   diffq:=(f(x+h)-f(x))/h;
```
$$diffq := \frac{(x+h)^3 - 8h - x^3}{h}$$
```
>   simplify(diffq);
```
$$3x^2 + 3xh + h^2 - 8$$

Notice that the derivative, $3x^2 - 8$, is obtained by letting h tend to zero (any term containing an h will disappear).

4.2 Differentiating Functions

All the rules for differentiation are programmed into Maple, making it easy to differentiate complicated functions. The Maple syntax for differentiating functions is different from that for expressions. So read carefully!

Suppose that f is defined as a Maple function. Then D(f) is the *function* that represents the derivative of f.

EXAMPLE 1:   Enter $f(x) = x^2(x^5 + 1)$ as a function and compute $f'(2)$.
SOLUTION: The function is
```
>   f:=x->x^2*(x^5+1);
```
$$f := x \to x^2 (x^5 + 1)$$

and the derivative is
```
>   D(f);
```
$$x \to 2x(x^5 + 1) + 5x^6$$

The arrow notation indicates that D(f) is a function. To evaluate $f'(2)$, type
```
>   D(f)(2);
```
$$452$$

If you are going to use the derivative repeatedly, it is best to give it a name. Thus in
```
>   Df:=D(f);
```

4.2. DIFFERENTIATING FUNCTIONS

$$Df := x \to 2x(x^5 + 1) + 5x^6$$

D(f) is the operation of taking the derivative of f, while Df is the name for the result. Then the derivatives of f at the points $x = -3$, $x = 3$, and $x = t$ are obtained by typing

>   Df(-3), Df(3), Df(t);

$$5097,\ 5109,\ 2t(t^5 + 1) + 5t^6$$

EXAMPLE 2: Find the tangent line at $x = 2$ for the function

>   f:=x->(x^3-1)/(x+2);

$$f := x \to \frac{x^3 - 1}{x + 2}$$

Then plot the function and its tangent line.

SOLUTION: The tangent line passes through the point $(2, f(2))$ and has a slope $f'(2)$, which in Maple is D(f)(2). First assign the slope to the variable m.

>   m:=D(f)(2);

$$m := \frac{41}{16}$$

Then the tangent line is given by

>   y:=m*(x-2)+f(2);

$$y := \frac{41x}{16} - \frac{27}{8}$$

A plot of the expressions $f(x)$ and y shows the tangent line has been found.

>   plot([f(x),y],x=0..4);

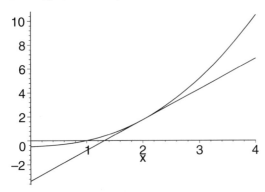

Higher derivatives can be evaluated by repeated application of the D operator. For example, D(D(f)) represents the second derivative of the function f (so $f''(2)$ can be computed by entering D(D(f))(2)). Alternatively, the syntax (D@@2)(f) also represents the second derivative. This syntax is preferred for higher derivatives; e.g., the 8th derivative, $f^{(8)}(2)$ is evaluated by typing

>   (D@@8)(f)(2);

$$\frac{-2835}{2048}$$

4.3 Differentiating Expressions

The `D` operator only works for functions. Sometimes, however, it is more convenient to work with expressions. To differentiate an expression, you first display the derivative using the `Diff` command. The first argument is the expression to be differentiated while the second argument is the variable of differentiation. This is followed by `value(%);` to compute the derivative.

EXAMPLE 1: Find the derivative of $f = \frac{3x^2+4x^3}{2x+1}$ and evaluate it at $x = 2$.

SOLUTION: The derivative is

> `Diff((3*x^2+4*x^3)/(2*x+1),x); Df:=value(%);`

$$\frac{d}{dx}\left(\frac{3x^2+4x^3}{2x+1}\right)$$

$$Df := \frac{6x+12x^2}{2x+1} - \frac{2(3x^2+4x^3)}{(2x+1)^2}$$

By first displaying the derivative, you can be sure it is entered correctly before differentiating. Notice that the derivative of `f` has been assigned to the variable `Df`. If you want to simplify the answer, execute

> `Df:=simplify(Df);`

$$Df := \frac{2x(8x^2+9x+3)}{(2x+1)^2}$$

To evaluate the derivative at $x = 2$, use the `subs` or `eval` command.

> `subs(x=2, Df);`

$$\frac{212}{25}$$

If the expression f has already been assigned to a label `f`, as is frequently the case, then it is more convenient to differentiate directly using the `diff` command.

EXAMPLE 2: Enter the function $\cos^3(t^2)$ as an expression and find its derivative at $t = \sqrt{\pi}/2$.

SOLUTION: In Maple the power on a trig function, such as the cube on $\cos^3(t^2)$, must be entered after the argument as follows:

> `f:=cos(t^2)^3;`

$$f := \cos(t^2)^3$$

Now differentiate f with respect to t and give it a label.

> `df:=diff(f,t);`

$$df := -6\cos(t^2)^2 \sin(t^2)\, t$$

Finally evaluate at $t = \sqrt{\pi}/2$:

> `eval(df, t=sqrt(Pi)/2);`

$$-\frac{3\sqrt{2}\sqrt{\pi}}{4}$$

NOTE: `eval` automatically simplifies. `subs` would not. Try it!

4.4. IMPLICIT DIFFERENTIATION

Higher derivatives can be calculated by repeating the variable of differentiation or by using a dollar sign followed by the number of derivatives to be taken. For example, the second and third derivatives of f are

> `d2f:=simplify(diff(f,t,t)); d3f:=(diff(f,t$3));`

$$d2f := -6\cos(t^2)\left(-4t^2 + 6\cos(t^2)^2 t^2 + \cos(t^2)\sin(t^2)\right)$$

$$d3f := -48\sin(t^2)^3 t^3 + 168\cos(t^2)^2 \sin(t^2) t^3 + 72\cos(t^2)\sin(t^2)^2 t - 36\cos(t^2)^3 t$$

If `f` is a function, its derivative is either the function `D(f)` or the expression `diff(f(x),x)`. (Here, you must type `f(x)` and not just `f`.) The disadvantage in the `diff` syntax is that you must use a more cumbersome `subs` or `eval` command to evaluate the derivative at particular points.

4.4 Implicit Differentiation

Up until now, we have discussed functions that are defined *explicitly*, which means that the dependent variable, such as y or f, is isolated on one side of an equation and an expression involving only the dependent variable, say x, appears on the other side (e.g., $y = x^2$). In this chapter, we consider functions that are given *implicitly*. This means that the variables are often mixed together in an equation.

An example is the Folium of Descartes, which is given by the equation $3xy = x^3 + y^3$. Note that it would be very difficult to solve for the variable y explicitly in terms of x since this is a cubic in y. Nevertheless, plots and derivatives can still be obtained.

To plot the equation, $3xy = x^3 + y^3$, use the `implicitplot` command from the `plots` package.

> `with(plots):`
> `eq:=3*x*y=x^3+y^3;`

$$eq := 3\,x\,y = x^3 + y^3$$

> `implicitplot(eq, x=-3..3, y=-3..3, scaling=constrained);`

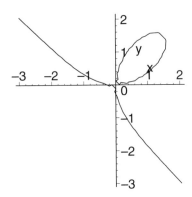

You can increase the number of points plotted by adding the option `grid=[p,q]`, where `p` and `q` are the numbers of points in the x and y directions. The default is `grid=[25,25]`.

```
>  implicitplot(eq, x=-3..3, y=-3..3, scaling=constrained,
>  grid=[201,201]);
```

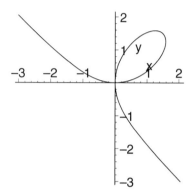

Note that this plot contains a loop, and cannot be described globally as the graph of one function $y = y(x)$. However, near most points a piece of the plot is the graph of one function. For example, the lower piece of the loop over the interval $-1 \leq x \leq 1$ is the graph of one function $y(x)$. Finding a formula for $y(x)$ involves solving the equation $3xy = x^3 + y^3$ for y in terms of x. This is difficult since this equation involves a cubic. However, it is possible to find numerical values of $y(x)$ at specific values of x using `fsolve`. For example, the three values of y at $x = 1$ are found using:

```
>  eq1:=subs(x=1, eq);
```
$$eq1 := 3\,y = 1 + y^3$$

```
>  fsolve(eq1,y);
```
$$-1.879385242,\ 0.3472963553,\ 1.532088886$$

What about the points where $y = x$? They can be found exactly:

```
>  solve({y=x,eq},{x,y});
```
$$\{x = 0,\ y = 0\},\ \{x = 0,\ y = 0\},\ \{x = \frac{3}{2},\ y = \frac{3}{2}\}$$

Near the point $(x, y) = (1.5, 1.5)$ the graph is actually a function, as can be seen from an enlarged plot that limits the range of x and y.

```
>  implicitplot(eq, x=1.25..1.75, y=1.25..1.75, scaling=constrained);
```

4.4. IMPLICIT DIFFERENTIATION

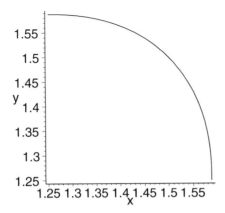

Indeed, over a very small plot range, the graph usually looks like a straight line (the tangent line).

> `implicitplot(eq, x=1.49..1.51, y=1.49..1.51, scaling=constrained);`

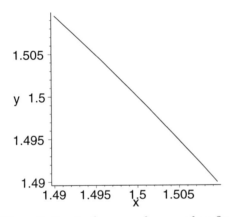

Implicit differentiation is the procedure used to find the derivative of an implicitly defined function. We will demonstrate the sequence of Maple commands used to find implicit derivatives on the Folium of Descartes.

First, assign the label `eq` to the equation. Then replace y by $y(x)$ (the unknown function).

> `eq:=3*x*y=x^3+y^3; eq1:=subs(y=y(x),eq);`

$$eq := 3xy = x^3 + y^3$$
$$eq1 := 3xy(x) = x^3 + y(x)^3$$

Next, take the derivative of both sides of the equation. Each side of this equation is an expression in x, so the `diff` command must be used to differentiate it. The left side requires the product rule; the right side requires the chain rule.

> `Deq:=diff(eq1,x);`

$$Deq := 3y(x) + 3x\left(\tfrac{d}{dx}y(x)\right) = 3x^2 + 3y(x)^2\left(\tfrac{d}{dx}y(x)\right)$$

Then solve for the derivative $\frac{d}{dx}y(x)$ and label it `Dy`:

```
> Dy:=solve(Deq,diff(y(x),x));
```

$$Dy := \frac{y(x) - x^2}{-x + y(x)^2}$$

There is no further need to emphasize that y depends on x; so substitute y back for $y(x)$: (Simplify if necessary.)

```
> Dy:=eval(Dy,y(x)=y);
```

$$Dy := \frac{y - x^2}{-x + y^2}$$

The expression Dy is the *implicit derivative*, and its numerical value can be determined at any point (x, y) by inserting the x and y values into its formula. For instance, the slope at the point $(3/2, 3/2)$ is

```
> m:=subs({x=3/2,y=3/2},Dy);
```

$$m := -1$$

This result agrees with the preceding graph.

4.5 Linear Approximation

In the previous section, we studied the Folium of Descartes,

```
> eq:=3*x*y=x^3+y^3;
```

$$eq := 3\,x\,y = x^3 + y^3$$

which implicitly defines a function $y = f(x)$ in the neighborhood of the point $\left(\frac{3}{2}, \frac{3}{2}\right)$. We found the slope at this point is $f'\left(\frac{3}{2}\right) = -1$.

```
> (a,b):=(3/2,3/2); m:=-1;
```

$$a, b := \frac{3}{2}, \frac{3}{2}$$

$$m := -1$$

So the tangent line is

```
> taneq:=y=m*(x-a)+b;
```

$$taneq := y = -x + 3$$

We can convert the right hand side of this equation into a function by using the `rhs` and `unapply` commands:

```
> ftan:=unapply(rhs(taneq),x);
```

$$ftan := x \rightarrow -x + 3$$

This tangent line function represents the linear approximation to the graph of the original function, i.e. the linear function $y = f_{tan}(x)$ approximates the values of $y = f(x)$ for x near 1.5. To plot both the curve and its tangent line, use the `implicitplot` command.

```
>  with(plots):
>  implicitplot([eq,taneq], x=1..2, y=1..2, scaling=constrained);
```

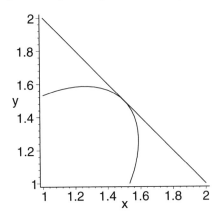

Let's compare the values of $y = f(x)$ and $y = f_{tan}(x)$ at $x = 1.51$. To find the value of $f(x)$ at $x = 1.51$, use `subs` and `fsolve`.

```
>  subs(x=1.51,eq); fsolve(%,y,y=1.4..1.6);
```

$$4.53\,y = 3.442951 + y^3$$
$$1.489436032$$

On the other hand, the value from the linear approximation is

```
>  ftan(1.51);
```

$$1.49$$

These are very close.

4.6 Summary

Differentiation of Functions and Expressions

- Compute a derivative (i.e., the slope of a tangent line) by finding the limit of the difference quotient. Here are the steps:

  ```
  f:=x-> expr;               Define the function.
  Limit((f(a+h)-f(a))/h,h=0);  Compute the limit of the
    m:=value(%);             difference quotient.
  ```

- Compute the line tangent to the graph of $y = f(x)$. Display both the curve and its tangent line on the same graph.

  ```
  y:=m*(x-a)+f(a);          Define the tangent line.
  plot([y,f(x)],x=0..4);    Plot the curve and the tangent line.
  ```

- Compute the derivative of a Maple function f, and evaluate it.

  ```
  Df:=D(f)                  Compute f'(x).
  Df(a);                    Evaluate f'(a).
  ```

- Compute a higher derivative of a Maple function f, and evaluate it.

`Dnf:=(D@@n)(f)`	Compute $f^{(n)}(x)$.
`Dnf(a);`	Evaluate $f^{(n)}(a)$.

- Compute the derivative of a Maple expression f, and evaluate it.

`Df:=diff(f,x)`	Compute $f'(x)$.
`subs(x=a,Df);`	Evaluate $f'(a)$.

- Compute a higher derivative of a Maple expression f, and evaluate it.

`Dnf:=diff(f,x$n)`	Compute $f^{(n)}(x)$.
`subs(x=a,Dnf);`	Evaluate $f^{(n)}(a)$.

Implicit Differentiation

- Maple expressions and functions explicitly define functions, $y = f(x)$.

- Maple equations implicitly define functions, $y = f(x)$, along pieces of their graphs. One equation may define several implicit functions.

- Plot a Maple equation.

`with(plots);`	Load the `plots` package.
`eq:=`*equation*`;`	Define the equation.
`implicitplot(eq, x=p..q, y=r..s);`	Plot the equation.

- The implicit derivative is the derivative of an implicit function. Maple can find this derivative, even if it cannot solve for y.

- Compute an implicit derivative and evaluate at a point.

`eq1:=subs(y=y(x),eq);`	Express y as a function of x.
`Deq:=diff(eq1,x);`	Differentiate the equation.
`Dy:=solve(Deq,diff(y(x),x));`	Solve for the derivative.
`Dy:=subs(y(x)=y,Dy);`	Suppress the dependence on x.
`Dy:=simplify(Dy);`	Simplify if necessary.
`m:=subs(x=a,y=b,Dy);`	Evaluate at a point.

Linear Approximation

- The linear approximation to a curve at a point $(x, y) = (a, b)$ on the curve is the function given by the formula for the tangent line.

`taneq:=y=m*(x-a)+b;`	Equation of the tangent line.
`ftan:=unapply(rhs(taneq),x);`	Convert equation to function.
`with(plots);`	Load the `plots` package.
`implicitplot({eq,taneq], x=p..q, y=r..s);`	Plot the function and tangent.

4.7 Exercises

1. Define each of the following as a Maple function. Find the slope of the secant line between the two points. NOTE: θ is entered as theta.

   (a) $f(x) = \dfrac{x^2 - x}{x^2 + 1}$ between $x = 2$ and $x = 4$

   (b) $f(\theta) = \cos(\theta)$ between $\theta = \pi/6$ and $\theta = \pi/3$

   (c) $f(x) = 3^x$ between $x = 0$ and $x = 2$

2. Define each of the following as a Maple function. Find the slope of the secant line between $x = a$ and $x = a + h$. Simplify the answer using expand, factor and/or simplify. Find the slope of the tangent line by taking the limit of the slope of the secant line as h approaches 0.

   (a) $f(x) = \dfrac{x^2 - x}{x^2 + 1}$ for $a = 2$

   (b) $f(\theta) = \cos(\theta)$ for $a = \pi/6$

   (c) $f(x) = 3^x$ for $a = 0$

3. Define each of the following as a Maple function. Find the derivative using D and evaluate at a.

   (a) $f(x) = \dfrac{x^2 - x}{x^2 + 1}$ for $a = 2$

   (b) $g(t) = \cos^2(t^3 + 1)$ for $a = 1$

   (c) $r(\theta) = \theta^2 + \sin(\theta)\cos(\theta)$ for $a = \pi/3$

4. Repeat Exercise 3 but define the given functions as expressions and then use diff to take the derivatives.

5. Define $f = \sin(px^3) + \cos(qx^2)$ as a function of x (with p and q constant) and differentiate with respect to x. Now define f as a function of q (with x and p constant) and differentiate with respect to q.

6. Repeat Exercise 5 but define the function as an expression and use diff to take the derivative.

7. Define $f(x) = 2x^4 - 3x^2$ as a Maple function. Find the equation of the tangent line at the point $x = 2.3$. Plot both the function and the tangent line on the same coordinate axes, with different colors.

8. Repeat Exercise 7 but define the function as an expression.

9. Find the (angle of) inclination of the tangent line in Exercise 7 in degrees. HINT: Recall that the tangent of the inclination is the slope of the line. So this problem amounts to taking the inverse tangent of the slope of the line in Exercise 7. In Maple, the inverse tangent, arctan, gives radians.

10. Plot the line $y = 1 - 3x$ and the curve $y = \sqrt{34.6 - x^2}$. (In Maple, enter the $\sqrt{u}$ as sqrt(u).) Find their point of intersection. Then find the acute angle between the tangent lines at this point of intersection. Look at the plot to determine if you need to add or subtract their inclinations.

11. Define $f(x) = \sin(x) + 2\cos(x)$ and $g(x) = x^3 - 2x + 4$ as Maple functions. Then use D to differentiate $f(x)g(x)$, $f(x)/g(x)$, $(f \circ g)(x)$ and $(g \circ f)(x)$. (For Maple functions, the product fg is entered as f*g, the quotient f/g is entered as f/g and the composition $f \circ g$ is entered as f@g.) Evaluate these functions at x to get the expressions.
    NOTE: The derivatives written as functions are very cryptic.

12. For each equation, use implicit differentiation to find the slope at the indicated point. Then plot the equation and the tangent line at the indicated point.

    (a) $x^4 + x^2 y^2 + y^4 = 48$ at $(x, y) = (2, 2)$
    (b) $(x^2 + y^2 - 2x)^2 = 16x^2 + 12y^2$ at $(x, y) = (2, 4)$
    (c) $x^2 y + xy^2 = 30$ at $(x, y) = (3, 2)$

13. For each equation, use implicitplot to plot the given equation. Then find the y-values of the upper piece of its plot at $x = 1$, 1.25, 1.5, 1.75 and 2. Compute the slopes of the tangent lines at $x = 1$ and $x = 1.25$ by implicit differentiation and compare these values to the slope of the secant between $x = 1$ and $x = 1.25$.

    (a) $(x^2 + y^2)^2 = 9(x^2 - y^2)$
    (b) $x^2 y + xy^2 = 25$

14. Graph the equations $y = x^2$ and $x^4 - 3x + 2y^2 + 6y = 14$. Be sure to use scaling=constrained. Find the point(s) of intersection. For each point of intersection, find the acute angle between the tangent lines of both equations.

15. By following the steps given below, show that the derivative obtained implicitly (without solving for y) is the same as the derivative obtained explicitly by solving for y.

    (a) Consider the equation $x^2 + xy + y^2 = 81$. Use the procedure described in the text to compute the implicit derivative and label it Dy.
    (b) Use Maple to solve the equation for y and label the result as sol.
    (c) Execute Dy1:=subs(y=sol[1],Dy); to replace y by its explicit form.
    (d) Execute Dy2:=diff(sol[1],x); to take the derivative explicitly.
    (e) Compare your answers in (c) and (d) by subtracting.
    (f) Repeat your work with sol[2].

4.7. EXERCISES

16. One of the virtues of the implicit derivative process is that, given an equation relating x and y, it is not necessary to explicitly solve for y in order to compute y'. On the other hand, the answer is sometimes not as nice as an explicit answer.

    (a) Enter the equation $3y^2 - 7xy + 2x^2 = 0$ and label it eq. Use the procedure described in the text to compute the implicit derivative.

    (b) Now use with(plots): and implicitplot to have Maple draw the graph. (Increase the grid size.) Give a better answer for the derivative than the one you got in (a).

    (c) Try factor(eq);. Now explain why the answer in (b) is true.

    (d) Without executing any Maple commands, use your *insight* from the graph and factor to give the derivative at any point on the graph, except the origin. What is the derivative at the origin?

17. In the case of the Folium of Descartes, $3xy = x^3 + y^3$, for *most* points (x, y) on the graph there is a plot range for which the graph is a function. However, there are two exceptions where there are vertical tangents.

    (a) After plotting with the implicitplot command, click with your mouse to find approximate floating point decimal coordinates for the two points on the graph that have vertical tangents.

    (b) Refine your guess as to the coordinates of these two points by noting that the denominator of y' is $x - y^2$. The tangent is vertical when $x - y^2 = 0$. Use fsolve to find a better approximation. You may need to limit the range of the x- and y-variables as in:
    > fsolve({x-y^2=0,eq}, {x=p..q,y=r..s});

    (c) To find the *exact* x- and y-coordinates of the two points.use
    > solve({x-y^2=0,eq},{x,y});

    The answer is given in terms of RootOf's, as discussed in Section 2.3. Resolve the RootOf's using allvalues. Confirm the real solutions agree with your approximate solutions by applying evalf to the exact answers.

18. An object's position at time t is $x(t) = 481t - 183t^2 + 24t^3 - t^4$. Find all times T when the instantaneous velocity of the object (its derivative) equals the average velocity of the object over the interval $0 \leq t \leq T$. (There are 3 solutions, but ignore the one at $T = 0$.) Then plot $x(t)$ and the two tangent lines on one plot. (NOTE: The tangent line at $T = a$ is tan1:=x(a)+Dx(a)*(t-a);.) What do you observe and why does it happen?

19. *This exercise does not involve Maple. It will confirm your understanding of the answer in the previous problem and provide a warm-up for the next exercise, which does involve Maple. Suppose the curve pictured here*

represents a plot of an object's distance versus time. From this plot, determine the approximate time T when the instantaneous velocity of the object equals the average velocity of the object over the interval $0 \leq t \leq T$.

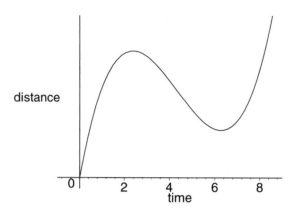

20. Consider the following data, which represent the position (in meters) of an object at various times (in seconds).

Time	1	2	3	4	5	6	7	8	9	10
Pos.	0.80	1.34	1.72	2.06	2.21	2.49	2.81	3.41	4.19	4.64

Plot this data set using Maple. From this plot, estimate the time(s) T at which the instantaneous velocity is equal to the average velocity over the time interval $0 \leq t \leq T$.

HINT: Enter the data as a list, with alternating times and positions, i.e., `data:=[[1,0.80],[2,1.24], ...];` as done in Section 3.4. Then type `plot(data);`. This will connect all the data points with line segments. To obtain a plot of data points without the connecting line segments, add the option `style=point`.

4.7. EXERCISES

21. *Newton's Method.* We have already seen examples of solving equations, such as $x^3 + x - 1 = 0$, numerically with the `fsolve` command. But how does Maple do it? The point of this exercise is to explore one algorithm, called Newton's method, which is often used to solve equations numerically. The basic idea behind Newton's method is as follows: (For more details, see the section on Newton's method in your text.) To solve the equation $f(x) = 0$, pick a starting point, x_0, near the solution to $f(x) = 0$ (for example, from a graph, choose the closest integer to the solution). Construct the tangent line to $y = f(x)$ at $x = x_0$. Generally, the x-intercept of the tangent line is closer to the solution than is x_0.

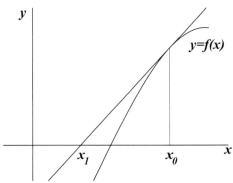

We denote this x-intercept by x_1. After computing the equation of the tangent line and finding its x-intercept, we obtain

$$x_1 = x_0 - \frac{f(x_0)}{f'(x_0)}.$$

Then we iterate this process. This leads to a sequence of points

$$x_{n+1} = x_n - \frac{f(x_n)}{f'(x_n)}$$

which approaches the solution to the equation $f(x) = 0$ as n gets bigger (in fact, n usually does not have to get too large). This algorithm can be made into a Maple command as follows:

```
> Df:=D(f);
> Newton:=x->evalf(x-f(x)/Df(x));
```

If x is taken as x_n then `Newton(x)` will be x_{n+1}. This command can then be used to solve $x^3 + x - 3 = 0$ using one of the following methods:

i. Enter the function and compute the derivative:

```
> f:=x->x^3+x-3; Df:=D(f);
```

Plot it and notice there is a solution of $f(x) = 0$ in the interval $1 \leq x \leq 2$. So start with $x_0 = 1$:

```
> x0:=1;
```

To get x_1, the first iteration of Newton's method, execute

```
> Newton(%);
```

Now repeatedly execute the command `Newton(%);` until the output no longer changes. (Use copy and paste to obtain multiple copies of the command.) Compare your answers with the result of Maple's `fsolve` command.

ii. You can also write a loop to execute Newton's method, say 5 times:

```
> x0:=1;
> for n from 1 to 5 do
> x[n]:=Newton(%)
> end do;
```

The commands between the `do` and the `end do;` will be executed 5 times (with the variable n as the counter). The value of x after this program is executed will be an approximation to the solution. Compare the loop with your result in part (a).

iii. The loop can be automated to execute Newton's method until the difference between successive approximations (i.e., $|x_{n+1} - x_n|$) is less than some pre-assigned small number, such as 10^{-12}. One way to do this is to keep track of the current value `xnew` and previous value `xold` and assign the variable *tol* (for tolerance) to the absolute value of their difference. The algorithm should continue to execute until *tol* is less than the pre-assigned value. To implement this in Maple, first define the desired tolerance and set the current tolerance to 1 so the loop can start. Also increase the number of `Digits`.

```
> desiretol:=10^(-12); tol:=1; Digits:=15;
```

Then the loop is

```
> xold:=1;
> for n from 1 to 100 while (tol>desiretol) do
> xnew:=Newton(xold);
> tol:=abs(xnew-xold):
> xold:=xnew:
> end do;
```

This syntax will execute the statements between the `do` and the `end do` as long as the variable *tol* exceeds 10^{-12} or until the counter n reaches 100. (The choice of 100 is arbitrary but is included to keep the program from running indefinitely if *tol* never gets below 10^{-12}.)

For this exercise, use Newton's method to find all solutions to each of the following equations to 25 decimal places. (Remember to set your `Digits` back to 10 afterwards.) Compare your answers with the result of `fsolve`.

(a) $x^3 - 13x - 7 = 0$ (3 solutions)

(b) $x^3 = 3x - 1.5$ (3 solutions)

(c) $4\cos(x) = 0.9x$ (3 solutions)

4.7. EXERCISES

22. A cylindrical can is to contain 800 cubic centimeters. Find the dimensions of the can if its surface area including the top and bottom is 500 square centimeters. HINT: Plot the equation you need to solve. Notice there are two answers. Use several iterations of Newton's method to solve the resulting equation near each solution. Compare your answers to the ones obtained by using `fsolve`.

23. An waffle cone is to contain 51 cubic centimeters of ice cream. Find the dimensions (radius and height) of the cone if its lateral surface area is 71 square centimeters. Use Newton's method to solve the equation. Compare your answers to the ones obtained by using `fsolve`. Notice there are two solutions. Which one is real world? NOTE: The lateral surface area of a cone is $A = \pi r \sqrt{r^2 + h^2}$

24. The bottom of a trough is constructed from a 4 foot by 6 foot rectangular piece of metal by bending it so that the 4 foot width forms an arc of a circle. (See the figure)

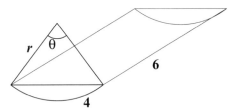

If the volume of the trough is 5 cubic feet, find the angle θ subtended by the arc. Use several iterations of Newton's method to solve the resulting equation for the volume (see the hint below) and compare your answer to the one obtained by using `fsolve`.

HINT: Suppose that r is the radius of the circular arc subtended by the angle θ. Explain why the length of the arc is $L = r\theta = 4$ and why the area of an end of the trough is

$$A = \frac{r^2 \theta}{2} - \frac{r^2 \sin(\theta)}{2}$$

(Think of the area of the end as the area of the circular sector minus the area of a triangle.) Use these equations to show the volume of the trough is:

$$V = \frac{48(\theta - \sin(\theta))}{\theta^2}$$

25. A pulley consists of an 40 inch band tautly wrapped around two wheels of radius 3 inch and 4 inches, respectively, as shown in the diagram. Find the length of the straight pieces of the band that are not in contact with either wheel. Use several iterations of Newton's method to solve the equation and compare your answer to the one obtained by using `fsolve`.

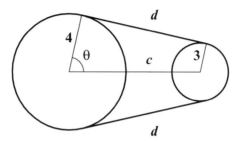

HINT: Let c be the distance between the centers and let θ be the angle between the line between their centers and a radius to the point where the band makes contact with the wheel. The two unknowns are θ and c. The two equations express (i) the total length of the band and (ii) the fact that the radii to the points of contact are perpendicular to the piece of band not in contact. Express the points of contact in terms of θ and c. Use the facts that the length of an arc is $l = r\theta$ and two lines are perpendicular if their slopes are negative reciprocals. Solve for c exactly from one equation and substitute into the other.

Chapter 5

Applications of Differentiation

Maple is used in this chapter to solve some classic problems associated with differentiation; namely, related rate problems, graphical analysis problems and max/min problems.

5.1 Related Rates

In many situations, two or more quantities are related by some formula and are changing with time. In a related rates problem, the goal is to determine how fast one of the quantities is changing when the rates of change of the other quantities are known.

The procedure for solving this type of problem can be outlined as follows:

1. Determine the quantities that are changing with time (or the parameter).

2. Find an equation that relates the quantities (this may involve known formulas from physics or geometry).

3. Enter the equation into Maple, writing the quantities as functions of time.

4. Take the implicit derivative of the equation with respect to time.

5. Solve the resulting equation for the desired derivative.

6. Find numeric values for each of the quantities in the resulting solution, possibly by using the equation from step 3, and substitute them into the result.

The following is a typical example.

EXAMPLE: For a concave lens with focal length 30 cm, the following formula from optics expresses the relationship between the distance u from an object to the lens and the distance v from the lens to its image.

$$\frac{1}{30} = \frac{1}{u} + \frac{1}{v}$$

Suppose an object is moving towards the lens at a rate of 3 cm/sec. Find the rate at which the image is receding from the lens when the object is 90 cm away.

SOLUTION: The two quantities that change with time are the two distances, u and v. The equation that relates them is the optics formula given in the problem. This equation is entered into Maple, using $u(t)$ and $v(t)$ to signify that the distances are time-dependent.

> `eq:=1/30=1/u(t)+1/v(t);`

$$eq := \frac{1}{30} = \frac{1}{\text{u}(t)} + \frac{1}{\text{v}(t)}$$

This equation is differentiated with respect to t.

> `deq:=diff(eq,t);`

$$deq := 0 = -\frac{\frac{d}{dt}\text{u}(t)}{\text{u}(t)^2} - \frac{\frac{d}{dt}\text{v}(t)}{\text{v}(t)^2}$$

The resulting equation is solved for $v'(t)$.

> `vrate:=solve(deq,diff(v(t),t));`

$$vrate := -\frac{(\frac{d}{dt}\text{u}(t))\,\text{v}(t)^2}{\text{u}(t)^2}$$

The numerical value of v is found.

> `u0:=90; urate0:=-3;`

$$u0 := 90$$
$$urate0 := -3$$

> `solve(eq,v(t)); v0:=subs(u(t)=u0,%);`

$$\frac{30\,\text{u}(t)}{\text{u}(t) - 30}$$
$$v0 := 45$$

The answer is found by substituting these values into `vrate`.

> `vrate0:=subs({u(t)=u0, diff(u(t),t)=urate0, v(t)=v0}, vrate);`

$$vrate0 := \frac{3}{4}$$

So the image is moving away at $\frac{3}{4}$ cm/sec.

5.2 Local Extrema

In this section, you will learn how to use the `plot`, `diff`, `solve` and `fsolve` commands to find local maxima and minima (also called local extrema) of differentiable functions. If f is a differentiable function on an open interval, then its derivative must vanish (i.e., $f' = 0$) at each local maximum or minimum. So the strategy is to first plot the function to get an approximate idea of the location of the maxima and minima. Then use the `solve` or `fsolve` command to find the solutions of the equation $f' = 0$.

EXAMPLE: Plot the expression $f = x^3 + 0.2x^2 - x$ and find the location of the local maxima and minima.

SOLUTION: Define f and set Df equal to its derivative.

> `f:=x^3+0.2*x^2-x;`
$$f := x^3 + 0.2\, x^2 - x$$

> `Df:=diff(f,x);`
$$Df := 3\, x^2 + 0.4\, x - 1$$

Plot f for $-2 \leq x \leq 2$.

> `plot(f, x=-2..2);`

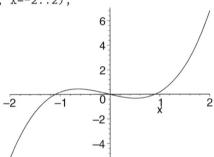

Observe there is a local maximum between -1 and 0 and a local minimum between 0 and 1. To find the precise locations, solve the equation $Df = 0$. The local maximum is between -1 and 0 and is labeled `xmax`:

> `xmax:=fsolve(Df=0,x=-1..0);`
$$xmax := -0.6478531925$$

The local minimum is between 0 and 1 and is labeled `xmin`.

> `xmin:=fsolve(Df=0,x=0..1);`
$$xmin := 0.5145198591$$

Inserting these values back into the original expression f will give the corresponding y-coordinates of the local maximum and local minimum.

> `ymax:=subs(x=xmax,f); ymin:=subs(x=xmin,f);`
$$ymax := 0.4598830456$$
$$ymin := -0.3253645270$$

Therefore, the local maximum is at the point $(-.6478531925, .4598830456)$ and the local minimum is at $(.5145198591, -.3253645270)$.

NOTE: Above f was defined as a function and we used `subs`. If instead, f is defined as an arrow-defined function,

```
> f:=x->x^3+0.2*x^2-x;
```
$$f := x \to x^3 + 0.2\, x^2 - x$$

then the x-coordinate of the local maximum can be found from

```
> xmax:=fsolve(D(f)(x)=0,x=-1..0);
```
$$xmax := -0.6478531925$$

and the corresponding y-coordinate can then be obtained from

```
> ymax:=f(xmax);
```
$$ymax := 0.4598830456$$

Of course, the above example could have been done by hand. However, the example in the next section, and many of the exercises are too complicated for hand computation.

5.3 Graphical Analysis

In this section, we consider, in detail, the graph of the expression:

$$f := \frac{e^x}{x^3 + x - 1 + 0.2 e^x}$$

The goal is to produce an accurate graph, locating horizontal and vertical asymptotes, local extrema and inflection points.

First, enter the expression into Maple with e^x entered as `exp(x)`.

```
> f:=exp(x)/(x^3+x-1+0.2*exp(x));
```
$$f := \frac{e^x}{x^3 + x - 1 + 0.2\, e^x}$$

Plot the graph over the interval $-5 \le x \le 15$.

```
> plot(f, x=-5..15, y=-10..10, discont=true);
```

The y-range is restricted to $-10 \le y \le 10$ to obtain a reasonable plot and a line at the vertical asymptote has been eliminated by including the `discont=true` option. From the graph, it appears that there is a vertical asymptote between $x = 0$ and $x = 1$. Its location can be pinpointed by locating the root of the

5.3. GRAPHICAL ANALYSIS

denominator of f. In Maple, the denominator of an expression f is `denom(f)`. Using `denom(f)` saves typing in cases where the denominator is complicated.

> `vasymp:=fsolve(denom(f)=0,x=0..1);`

$$vasymp := 0.5213890506$$

We find horizontal asymptotes by taking limits

> `Limit(f,x=-infinity); value(%);`

$$\lim_{x \to (-\infty)} \frac{e^x}{x^3 + x - 1 + 0.2\, e^x}$$

$$0.$$

> `Limit(f,x=infinity); value(%);`

$$\lim_{x \to \infty} \frac{e^x}{x^3 + x - 1 + 0.2\, e^x}$$

$$5.$$

So the lines $y = 0$ and $y = 5$ are horizontal asymptotes as $x \to -\infty$ and $x \to \infty$, respectively.

From the plot, there appears to be a local minimum between $x = 1$ and $x = 5$. Its location and value can be pinpointed by setting the derivative to zero.

> `Df:=diff(f,x):`
> `xmin:=fsolve(Df=0,x=1..5);`

$$xmin := 2.893289196$$

> `ymin:=evalf(subs(x=xmin,f));`

$$ymin := 0.6073428968$$

Try the last command without the `evalf`. Notice that Maple does not automatically simplify after a `subs` command.

It is instructive to replot both f and its derivative on the same coordinate axes. We focus on the part of the graph of f over the interval $0 \leq x \leq 15$, since the local minimum of f is located there.

> `plot({f,Df}, x=0..15, y=-2..6, discont=true);`

The scale for the y-coordinate is changed to more clearly display the graphs. Note that f decreases on the interval $0.53 < x < 2.89$ and the derivative of f

is negative on this same interval. Likewise, f increases on $x > 2.89$ and the derivative of f is positive on this same interval. The graph of f flattens as x gets large. This corresponds to the observation that the graph of the derivative approaches $y = 0$ when x gets large.

There also appears to be an inflection point (a point where the concavity of the graph switches) between $x = 6$ and $x = 10$. Its location can be pinpointed by computing the second derivative (the derivative of the first derivative, Df) and finding where it is zero.

```
>  DDf:=diff(Df,x):
>  xinfl:=fsolve(DDf=0,x=6..10);
```
$$xinfl := 8.131398912$$
```
>  yinfl:=evalf(subs(x=xinfl,f));
```
$$yinfl := 2.775853638$$

So the point of inflection is approximately $(8.131, 2.776)$.

We now plot the expression f and its second derivative on the same coordinate axes.

```
>  plot({f,DDf}, x=0..15, y=-1..5, discont=true);
```

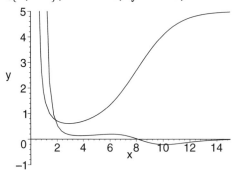

Notice that f is concave up and the second derivative is positive on the interval $0.53 < x < 8.13$. The graph of f is concave down and the second derivative is negative on the interval $x > 8.13$.

Also note that the graph of f is steepest where the first derivative is largest (see previous graph). This point, $x = 8.13$, is where the second derivative is zero.

5.4 Designer Polynomials

Here is an example that combines many of the Maple commands introduced up to this point. It is an extension of the ideas introduced in Example 1 of Section 3.1.

EXAMPLE: Find the coefficients of the cubic polynomial

$$f(x) = ax^3 + bx^2 + cx + d$$

5.4. DESIGNER POLYNOMIALS

that has a relative minimum at $(-2, 0)$ and a relative maximum at $(3, 4)$. Plot the points and the polynomial.

SOLUTION: Four equations are needed for the four unknowns a, b, c and d. These equations will be labeled eq1 through eq4 for later reference. After defining f as a Maple function, the first two equations can be obtained from the information $f(-2) = 0$ and $f(3) = 4$.

> f:=x->a*x^3+b*x^2+c*x+d;
$$f := x \to a x^3 + b x^2 + c x + d$$

> eq1:=f(-2)=0; eq2:=f(3)=4;
$$eq1 := -8a + 4b - 2c + d = 0$$
$$eq2 := 27a + 9b + 3c + d = 4$$

The remaining two equations can be obtained from the fact that the derivative of f must vanish at $x = -2$ and $x = 3$ (because f has a local minimum and maximum at these points).

> eq3:=D(f)(-2)=0; eq4:=D(f)(3)=0;
$$eq3 := 12a - 4b + c = 0$$
$$eq4 := 27a + 6b + c = 0$$

Now these four equations are solved for a, b, c, and d.

> sol:=solve({eq1,eq2,eq3,eq4},{a,b,c,d});
$$sol := \{a = \frac{-8}{125}, b = \frac{12}{125}, c = \frac{144}{125}, d = \frac{176}{125}\}$$

NOTE: In the solve command, Maple requires curly braces { } or square brackets [ ] enclosing the equations {eq1,eq2,eq3,eq4} and the variables {a,b,c,d}. If the variables are enclosed in braces, then the output sol is also in the form of a set. Within the set, your solutions may appear in a different order. If the variables are enclosed in brackets, then the output sol is a list of lists and must be handled differently (which will not be discussed here).

To substitute the values of a,b,c,d back into the original function, type

> g:=subs(sol,f(x));
$$g := -\frac{8}{125} x^3 + \frac{12}{125} x^2 + \frac{144}{125} x + \frac{176}{125}$$

Finally, we plot the points and the graph of g to see it has a local minimum at $(-2, 0)$ and a local maximum at $(3, 4)$.

> with(plots):

> p1:=plot([[-2,0],[3,4]], x=-4..5, style=point, symbol=box,
> symbolsize=24):

> p2:=plot(g, x=-4..5):

> display({p1,p2});

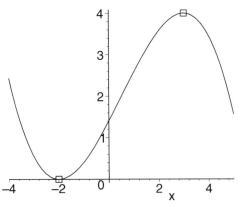

5.5 Absolute Extrema on an Interval

In this section, Maple is used to help find the absolute maximum and/or minimum (also called absolute extrema) of a function (or expression) on a (closed and bounded) interval. The method compares the values of the function at its critical points on the given interval with those at the endpoints of the interval. This is illustrated by the following example.

EXAMPLE 1: Find the absolute maximum and minimum of the expression $f = \ln(x) - 4x^2 + x^3$ on the interval $1 \le x \le 3$.

SOLUTION: First, enter f as a function in Maple and plot it over the interval $1 \le x \le 3$.

> `f:=x->ln(x)-4*x^2+x^3;`

$$f := x \to \ln(x) - 4x^2 + x^3$$

> `plot(f, 1..3);`

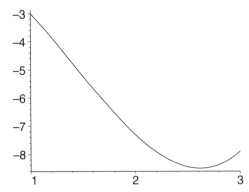

From the plot, it appears that the minimum of f is located at a critical point near $x = 2.6$ and that the maximum of f is located at the left hand endpoint $x = 1$. To find the critical point, set the derivative of f equal to zero and solve.

> `xmin:=fsolve(D(f)(x)=0,x=2..3);`

$$xmin := 2.618033989$$

The values of f at the critical point $x = xmin$ and the endpoints $x = 1$ and $x = 3$ are

> f(xmin), f(1), f(3.);
$$-8.50971230, -3, -7.90138771$$

So, the maximum of f is -3 at $x = 1$ and the minimum of f is -8.51 at $x = 2.62$.

If the interval is open (does not contain its endpoints) or if the interval is infinite, then there may not be an absolute maximum or minimum.

EXAMPLE 2: Find any absolute maxima or minima of the expression $f = \dfrac{x^4 + 2x + 3}{x^2}$ on the interval $x > 0$.

SOLUTION: Enter the expression and plot it.

> f:=(x^4+2*x+3)/x^2;
$$f := \frac{x^4 + 2x + 3}{x^2}$$

> plot(f, x=0..10, y=-50..100);

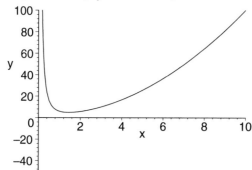

The plot reveals an absolute minimum at a critical point near $x = 1.5$. However, this expression has no absolute maximum, since f approaches infinity as $x \to 0^+$ or $x \to \infty$. The minimum can be found by setting the derivative, Df:=diff(f,x), equal to zero and solving: fsolve(Df=0,x=1..2);. The minimum of f is approximately 4.9087, occurring at $x = 1.4526$. Try it.

5.6 The Most Economical Tin Can

This section presents an example of an applied max/min problem that cannot be easily solved by hand. The discussion will focus on the problem solving strategy and the role of Maple in that strategy rather than on the Maple syntax itself.

For an applied max/min problem, you must

i. translate the problem into mathematics, (Maple will not help you here.)

ii. solve the resulting calculus problem, (This is where Maple can help.) and finally,

iii. interpret the results. (Again Maple cannot help.)

EXAMPLE: Consider a cylindrical tin can (shown below) which is to be constructed by joining the ends of a rectangular piece of material to form the curved side and then attaching circular pieces to form the top and bottom. There are seams around the perimeter of the top and bottom, and there is one seam down the side surface (where the ends of the rectangle join together).

Suppose the volume of the can is 750 cubic centimeters. Also suppose that the cost of the material is \$1.10 per square meter and the cost of the seam is \$0.25 per meter. Find the dimensions of the can that will minimize its cost. Find the minimum cost.

SOLUTION: Let r be the radius of the base of the can and h be the height of the can. The surface area of the can is

> `area:=2*Pi*r^2+2*Pi*r*h;`

$$area := 2\pi r^2 + 2\pi r h$$

The first term on the right represents the area of the top and bottom. The second term represents the area of the side (which has the same area as a rectangle of length $2\pi r$ and height h).

The total length of the seams is

> `len:=4*Pi*r+h;`

$$len := 4\pi r + h$$

We will use centimeters and cents as our units. The cost of the material is $a = 0.011$ cents per square centimeter (110 cents divided by the 10000 square centimeters in a square meter). The seaming cost is $b = 0.25$ cents per centimeter. So the total cost of the can is

> `a:=0.011; b:=0.25; cost:=a*area+b*len;`

$$a := 0.011$$
$$b := 0.25$$
$$cost := 0.022\pi r^2 + 0.022\pi r h + 1.00\pi r + 0.25 h$$

This cost expression has both r and h as variables. However, there is a constraint between the variables which we can use to eliminate h, namely that the volume is 750 cubic centimeters. Since the volume is

$$\text{volume} = \text{base} \times \text{height} = \pi r^2 h = 750$$

we have

> `h:=750/(Pi*r^2);`

5.7. SUMMARY

$$h := \frac{750}{\pi r^2}$$

The cost is now

> `cost;`

$$0.022\,\pi\,r^2 + \frac{16.500}{r} + 1.00\,\pi\,r + \frac{187.50}{\pi\,r^2}$$

which must be minimized over the interval $r > 0$. Its plot is

> `plot(cost,r=1..10);`

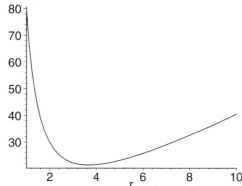

Observe that a minimum occurs somewhere between $r = 2$ and $r = 6$. To find it, set its derivative equal to zero and solve for r:

> `Dc:=diff(cost,r); rmin:=fsolve(Dc=0,r=2..6);`

$$Dc := 0.044\,\pi\,r - \frac{16.500}{r^2} + 1.00\,\pi - \frac{375.00}{\pi\,r^3}$$

$$rmin := 3.666752590$$

The answer is the best value for the radius (in centimeters). Substituting this value into the expression for h yields

> `hmin:=evalf(subs(r=rmin,h));`

$$hmin := 17.75612426$$

So the minimum cost (in cents) is:

> `cmin:=evalf(subs(r=rmin,cost));`

$$cmin := 21.38762546$$

5.7 Summary

Related Rates: The procedure for solving related rates problems can be outlined as follows:

1. Determine the quantities (say u and v) that are changing with time (or another parameter).

2. Find an equation that relates the quantities. (This may involve known formulas from geometry or physics, etc.)

3. Enter the equation (say eq) into Maple, writing the quantities as functions of time. (Here u(t) and v(t).)

4. Take the derivative of the equation with respect to time.
   ```
   deq:=diff(eq,t);
   ```

5. Solve the resulting equation for the desired implicit derivative.
   ```
   vrate:=solve(deq,diff(v(t),t));
   ```

6. Find numeric values for each of the quantities in the resulting solution, possibly by using the equation from step 3, and substitute them into the result.
   ```
   vrate0:=subs(u(t)=u0, diff(u(t),t)=urate0, v(t)=v0, vrate);
   ```

Graphical Analysis of Expressions: Use the following commands to analyze an expression labeled expr:

- Plot:
   ```
   >  f:=expr;
   >  plot(f,x);
   ```

- x-Intercepts: From the plot, determine an interval $[a, b]$ containing one x-intercept.
   ```
   >  fsolve(f=0,x=a..b);
   ```
   Repeat for each x-intercept.

- y-Intercept:
   ```
   >  subs(x=0,f);
   ```

- Local maxima and minima: From the plot determine an interval $[c, d]$ containing a maximum or minimum.
   ```
   >  Df:=diff(f,x);
   >  x0:=fsolve(Df=0,x=c..d);
   >  y0:=evalf(subs(x=x0,f));
   ```
   Repeat for each maximum or minimum.

- Inflection Points: From the plot determine an interval $[r, s]$ containing an inflection point.
   ```
   >  DDf:=diff(Df,x);
   >  xinfl:=fsolve(DDf=0,x=r..s);
   >  yinfl:=evalf(subs(x=xinfl,f));
   ```
   Repeat for each inflection point.

5.7. SUMMARY

- Vertical Asymptotes: From the plot determine an interval $[p, q]$ containing a vertical asymptote. If necessary, use `f:=simplify(f);` to write `f` as a single fraction, rather than a sum of fractions.
    ```
    > vasymp:=fsolve(denom(f)=0,x=p..q);
    ```
  Repeat for each vertical asymptote.

- Horizontal asymptote:
    ```
    > Limit(f,x=-infinity); value(%);
    > Limit(f,x=infinity); value(%);
    ```

Graphical Analysis of Arrow-Defined Functions: Use the following commands to analyze a function `f` defined from an expression `expr` by:
```
> f:=x->expr;
```

- Plot:
    ```
    > plot(f(x),x);
    ```

- x-Intercepts: From the plot, determine an interval $[a, b]$ containing one x-intercept.
    ```
    > fsolve(f(x)=0,x=a..b);
    ```
  Repeat for each x-intercept.

- y-Intercept:
    ```
    > f(0);
    ```

- Local maxima and minima: From the plot determine an interval $[c, d]$ containing a maximum or minimum.
    ```
    > Df:=D(f);
    > x0:=fsolve(Df(x)=0,x=c..d);
    > y0:=evalf(f(x0));
    ```
  Repeat for each maximum or minimum.

- Inflection Points: From the plot determine an interval $[r, s]$ containing an inflection point.
    ```
    > DDf:=D(Df);
    > xinfl:=fsolve(DDf(x)=0,x=r..s);
    > yinfl:=evalf(f(xinfl));
    ```
  Repeat for each inflection point.

- Vertical Asymptotes: From the plot determine an interval $[p, q]$ containing a vertical asymptote. If necessary, use `simplify(f(x));` to write `f(x)` as a single fraction, rather than a sum of fractions.
    ```
    > vasymp:=fsolve(denom(f(x))=0,x=p..q);
    ```
  Repeat for each vertical asymptote.

- Horizontal asymptote:
    ```
    >  Limit(f(x),x=-infinity); value(%);
    >  Limit(f(x),x=infinity); value(%);
    ```

Maximizing or Minimizing on an Interval:

- Define and plot the Maple function on the interval $a \leq x \leq b$.
    ```
    >  f:=x->expr;
    >  plot(f(x), x=a..b);
    ```

- Use the plot and the `fsolve` command to find the x-coordinates of the extrema in the interior of this interval.
    ```
    >  Df:=D(f);
    >  xcrit:=fsolve(Df(x)=0,x=c..d);
    ```

    If there is more than one critical point, repeat this and give them different labels.

- Evaluate the function at each of these critical points and the two endpoints:
    ```
    >  f(xcrit);
    >  f(a);
    >  f(b);
    ```

- Compare these. The largest is the absolute maximum; the smallest is the absolute minimum.

5.8 Exercises

1. When a ray of light hits the surface of a lake, the beam is bent. The equation that governs this effect is due to Willebrod Snell (1591–1626). He noted that
$$\frac{\sin(\alpha)}{\sin(\beta)} = 1.33$$
where 1.33 is the index of refraction of water (relative to air) and α and β are the angle of incidence and angle of refraction, respectively, measured from a line perpendicular to the surface of the lake. Since the earth rotates once in 24 hours, as the sun rises, the angle of incidence of the sunlight decreases at $\pi/12$ radians per hour. When $\alpha = \pi/3$, how fast does a fish see the sun rise?

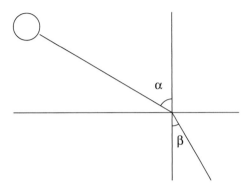

2. Plot the expression $f = .25x^4 - .31x^3 - .12x^2 + .25x - 0.4$. You may need to adjust the plot ranges to include all relevant aspects of the graph. Find the local maxima and minima and the intervals where f is increasing and decreasing.

3. Find the cubic $ax^3 + bx^2 + cx + d$ that has a local minimum at $(-1, -2)$ and a local maximum at $(4, 4)$. Plot it.

4. Find the cubic polynomial that passes through the points $(1, -1)$, $(5, 4)$, and $(6, 4)$ and whose derivative at $x = 3$ is 3.2. Plot it.

5. Find the cubic polynomial with a local maximum at $(2, 7)$ and a local minimum at $(5, 2)$. Plot it.

6. Plot one or more informative graphs of $y = \dfrac{|x^3 - 5.8x^2 + 11.5x - 5.4|}{x^3 - 3.1x^2 - 3.2x + 4.21}$. Determine all intercepts, asymptotes and local extrema.

7. Graph the function $y = \dfrac{x \ln(x)}{x^2 + x + +4}$ for $x > 0$. Find all critical points and inflection points. Find intervals where the function is increasing, decreasing, concave up, and concave down.

8. For each of the following expressions, locate the horizontal and vertical asymptotes, the local extrema and the inflection points. Then graph the given expression, its derivative and second derivative in one plot using different colors. On each of the intervals where the expression is increasing (decreasing), is the derivative positive or negative? Do a similar comparison for concavity of the expression and the sign of the second derivative.

   (a) $\dfrac{10x^2 + 4x + 1}{5x - 1}$

   (b) $\dfrac{3x^4 + 2x + 1}{x^4 - 3x^2 + 1}$

   (c) $\dfrac{3e^x}{e^x - x^6}$ (Type e^x as exp(x).)

(d) $x^2 + 4\sin(x)\cos(x)$

(e) $\ln(x) + 7x^2 - x^3$ for $x > 0$

9. (This is a warm-up for the next exercise.) A plot of land has straight southern, eastern and western boundaries and an irregular northern boundary. The goal of this problem is to draw a map of the boundary of this plot of land using cubic polynomials (called a cubic spline). Since the southern and western boundaries are straight, they will be represented by the x-axis and the y-axis respectively. North-south measurements (represented by the variable y) are taken and placed in the following table (each unit represents 100 feet).

x-value	0.0	1.0	2.0	3.0	4.0
y-value	3.2	1.5	2.6	3.2	2.7

The first task is to find the graph of the quadratic polynomial $p(x) = ax^2 + bx + c$ (a parabola) that contains the first three data points. This is easily accomplished by defining p as a function and solving the equations $p(0) = 3.2$, $p(1) = 1.5$, and $p(2) = 2.6$ for the unknowns a, b, and c.

The next task is to find the graph of the cubic polynomial $q(x) = dx^3 + ex^2 + fx + g$ that contains the last three data points and satisfies $p'(2) = q'(2)$. This last equation ensures that the graphs of p and q have the same slope at $x = 2$. To do this, enter q as a function (if necessary, unassign any previously used labels) and solve the equations $q(2) = 2.6$, $q(3) = 3.2$, $q(4) = 2.7$, and $D(p)(2) = D(q)(2)$ for d, e, f and g.

Since p and q are plotted over different intervals, graph them using the `display` command, as explained in Section 2.2

10. *Smooth off the southern boundary of Texas.* In Exercise 2 in Chapter 3, the boundary of the state of Texas was drawn using line segments. The point of this problem is to use parabolas and cubics (called a cubic spline) to smooth out the part of the southern boundary of the state formed by the Rio Grande River. Enter the relevant Rio Grande data:
```
> rio:=[[0,0], [1,-1.1], [2,-2.5], [3,-2.9], [4,-2.3],
>       [5,-2.8], [6,-4.4], [7,-5.8], [8,-6.1]]:
```

Here the origin is the westernmost corner of Texas (near El Paso) and the x-axis is the extension of the east-west border between New Mexico and Texas. Each unit represents approximately 69 miles.

To find functions that smooth out the Rio Grande, proceed as in Exercise 9. First, find the parabola $p(x)$ that passes through the three data points $[0, 0]$, $[1, -1.1]$, and $[2, -2.5]$. Then find a cubic polynomial q that passes through the next triplet of data points $[2, -2.5]$, $[3, -2.9]$, $[4, -2.3]$ and further satisfies the equation $p'(2) = q'(2)$ (so that the slopes of the graphs of p and q at $x = 2$ are the same). In the same manner, find cubics r and s for the triplets $[4, -2.3]$, $[5, -2.8]$, $[6, -4.4]$ and $[6, -4.4]$, $[7, -5.8]$, $[8, -6.1]$

5.8. EXERCISES

so that $q'(4) = r'(4)$ and $r'(6) = s'(6)$. Then `plot` $p(x)$ for $0 \le x \le 2$; $q(x)$ for $2 \le x \le 4$; $r(x)$ for $4 \le x \le 6$ and $s(x)$ for $6 \le x \le 8$ and `display` them on the same coordinate axes.

11. In Section 5.6 we found the dimensions of a tin can which minimized its cost. Repeat that problem, but assume you need to pay for each full square of material used to cut out the circular ends of the cylinder.

12. A metal box with a square base and no top holds 900 cubic centimeters. It is formed by folding up the sides of the flattened pattern pictured here and seaming up the four sides. The material for the box costs $1.00 per square meter and the cost to seam the sides is 6 cents per meter. Find the dimensions of the box that costs the least to produce assuming you do not need to pay for the metal squares which have been cut out of the corners.

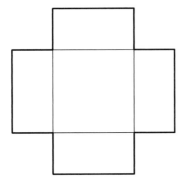

13. Repeat Exercise 12 but assume you do need to pay for the metal squares which have been cut out of the corners.

14. A pipeline is to be constructed to connect a station on the shore of a straight section of coast line to a drilling rig that lies 7 kilometers down the coast and 4 kilometers out at sea. Find the minimum cost to construct the pipeline, given that the pipeline costs 4.6 million dollars per kilometer to lay under water and 2.4 million dollars per kilometer to lay along shore.

15. A rectangular movie theater is 100 feet long (from the front screen to the back). The top and bottom of its screen are 40 feet and 15 feet from the floor, respectively. Find the position in the theater with the largest viewing angle.
    HINT: Arrange a coordinate system with the origin at the floor directly under the screen. You are asked to find the position x where the angle $\alpha - \beta$ is the largest (see figure). Instead of maximizing $\alpha - \beta$, it is easier to maximize $\tan(\alpha - \beta)$. This will lead to the same optimum value of x, since $\tan(\theta)$ is an increasing function on $-\pi/2 < \theta < \pi/2$. The subtraction formula for tangent will be needed. This may be found by executing `expand(tan(a-b));`.

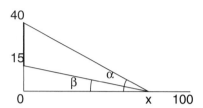

16. Repeat Exercise 15, but this time assume that the floor of the theater has a gentle parabolic slope given by the parabola $y = 0.001x^2$ for $0 \le x \le 100$. (The origin is located on the floor directly under the screen.)

17. Repeat Exercise 16, but assume that the floor of the theater has a steeper parabolic slope given by the formula $y = 0.004x^2$.

18. Two hallways of width a and b intersect at right angles. What is the length of the longest rigid rod that can be pushed on the floor around the corner of the intersection of these two halls? Carefully explain your steps. HINT: First try it for $a = 3$ and $b = 4$.

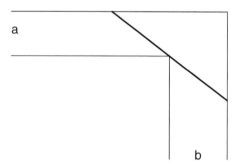

19. Find the point on the graph of $y = x^2 \ln(x)$ closest to the point $(2, 7)$. HINT: Minimize the function that describes the square of the distance between the point $(2, 7)$ and a typical point $(x, x^2 \ln(x))$ on the graph.

20. Find the point on the parabola $y = \dfrac{x^2}{3} - 1$ closest to the point $(-1, 5)$.

21. The point of this problem is to derive Snell's Law:

$$\frac{\sin(\theta_1)}{\sin(\theta_2)} = \frac{v_1}{v_2}$$

where θ_1 and v_1 are the angle of incidence and the velocity of light in the first medium (e.g. air), θ_2 and v_2 are the angle of refraction and the velocity of light in the second medium (e.g. water) as shown in the figure.

5.8. EXERCISES

(This equation can be used to calculate θ_2 provided θ_1, v_1, and v_2 are known.) We know from physics that a ray of light travels from a point A in the air to a point B in the water via a path ACB that minimizes the time taken where C is a point on the surface of the water.

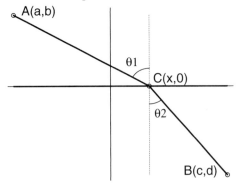

HINT: Write a formula for the time it takes to travel from A to C to B. Then vary C to minimize this formula. In the figure, a, b, c and d are fixed, x varies. Use Maple to compute a derivative, if you wish. Also express $\sin\theta_1$ and $\sin\theta_2$ in terms of a, b, c, d and x. The rest of the solution is more easily done by hand.

22. Now suppose the medium in Exercise 21 has a curved surface (such as the surface of a glass lens). To be more specific, suppose a beam of light follows the line with the equation $y = 3x + 2$ and strikes a piece of glass whose outside boundary is given by the equation $y = x^2$. (See the figure.) Find the equation of the line that represents the refracted light beam as it travels through the glass. Assume that v_1/v_2 is 1.5.

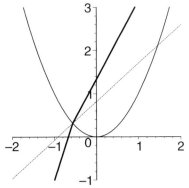

HINT: First, use Maple to find the point of intersection between the light beam and the parabolic surface of the glass. At this point compute the angle between the normal to the parabolic surface and the incoming light beam. This angle is the angle θ_1 of Exercise 21. Snell's law can now be used to calculate θ_2 and the slope of the refracted light can then be determined.

23. Suppose a glass lens is formed by the intersection of the interiors of the following two ellipses:

$$(x-4)^2 + \frac{y^2}{2} = 26 \quad \text{and} \quad (x+4)^2 + \frac{y^2}{2} = 26.$$

A light beam strikes this lens from the left along the horizontal line $y = 3$. Find the equations of the line segments that describe the trajectories of the refracted light beam as it passes through the lens and as it passes through the air on the right side of the lens. Find the location where the refracted light beam hits the positive x-axis.

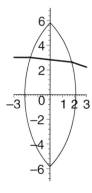

Chapter 6

Integrals

In the first section, we use Maple's `leftbox` and `rightbox` commands to visualize the process of approximating the area under a curve. In the second section, we compute some Riemann sums with the help of Maple's `Sum` command. The integration command `Int` is also introduced. The next three sections demonstrate how Maple can be used to perform integration by substitution, integration by parts, and integration using a partial fraction decomposition. The final section discusses the midpoint rule, the trapezoidal rule and Simpson's rule for approximating definite integrals.

Since Maple has a built-in integration command, it is reasonable to wonder why we need to discuss techniques of integration. It turns out that Maple cannot integrate everything, (Try integrating $\int x \ln(x + \sqrt{1+x^2})\,dx$.) but with a little help from us, the number of integrands that Maple is able to integrate in closed form increases. (By the end of the chapter you will compute $\int x \ln(x + \sqrt{1+x^2})\,dx$ in Exercise 14.)

6.1 Visualizing Riemann Sums

Let's compute the area of the region under the graph of the function $f(x) = x^2$ over the interval $1 \leq x \leq 3$. We enter this as the Maple function:

```
>    f:=x->x^2:
```

First, divide this interval into n subintervals of equal length, $\Delta x = \dfrac{3-1}{n} = \dfrac{2}{n}$. Here, n is typically a large number such as 10 or 100 (later, n will approach ∞). The area under the graph of $f(x) = x^2$ is approximated by the sum of the areas of n rectangles where the base of a typical rectangle is one of the n subintervals and the height is the value of the function $f(x)$ at the left or right endpoint of the subinterval. To get a picture with $n = 10$ rectangles, load the `student` and `plots` packages and enter the `leftbox` and `rightbox` commands as follows:

```
> with(student):  with(plots):
> Lplot:=leftbox(f(x), x=1..3, 10, xtickmarks=3,
> title="Left Riemann Sum"):

> Rplot:=rightbox(f(x), x=1..3, 10, xtickmarks=3,
> title="Right Riemann Sum"):

> display(array([Lplot,Rplot]));
```

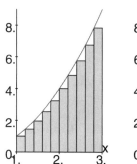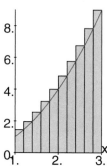

The `leftbox(f(x),x=1..3,10)` command draws the graph of $f(x)$ over the interval $1 \leq x \leq 3$, together with 10 rectangles; where the height of each rectangle is the value of the function at the left endpoint of the base of the rectangle. Similarly, the `rightbox` command takes the height of each rectangle as the value of the function at the right endpoint of the base. (The `display` command with an `array` argument shows the two plots next to each other instead of superimposing them.) The sum of the areas of these rectangles is called a left (right) Riemann sum approximation to the area under the graph of f.

Since the graph of f is increasing over the interval $1 \leq x \leq 3$, the left Riemann sum is less than the area under the graph of f (see the picture above) and is called a lower bound; while the right Riemann sum is more than the area under the graph and is called an upper bound.

The value $n = 10$ can be changed to any positive integer. Try the `leftbox` and `rightbox` commands with $n = 20$, 50 and 100. Note that, as n gets larger, the total area of the rectangles more closely approximates the area under the graph of f with the left Riemann sum as a lower estimate and the right Riemann sum as an upper estimate.

You can see this better with an animation. The `ApproximateInt` command in the `Student[Calculus1]` package allows you to animate Riemann sums. For example, to see a left Riemann sum with 4, 8, 16, 32 and 64 intervals execute

```
> with(Student[Calculus1]):

> ApproximateInt(x^2, 1..3, output=animation, iterations=5,
> method=left, partition=4, subpartition=all, refinement=halve);
```

6.2. THE COMPUTATION OF INTEGRALS

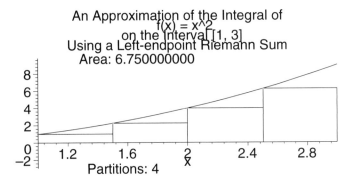

Try this yourself, since you cannot see an animation in a book. To animate it, click in the plot and click on the PLAY button which is the right pointing triangle (▷) on the plot toolbar. You can also access this animation as a Maplet accessable in the Standard interface from the menu TOOLS > TUTORS > CALCULUS - SINGLE VARIABLE > RIEMANN SUMS.

This same rectangle construction can be applied to any interval $a \leq x \leq b$. Try different values of a and b (and n). Note that, for negative values of a and b, f is a decreasing function on $a \leq x \leq b$. Therefore the left Riemann sums are upper estimates and the right Riemann sums are lower estimates.

6.2 The Computation of Integrals

We continue with the preceding example of the integral of $f(x) = x^2$ for $a \leq x \leq b$. and we look at a right Riemann sum.

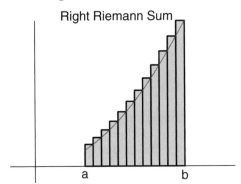

As in the previous section, the interval $a \leq x \leq b$ is divided into n subintervals, each of length

$$\Delta x = \frac{b-a}{n}$$

Let x_i denote the right endpoint of the ith subinterval, where i is a counting index that runs from 1 to n. The area of the ith box is the product of the interval length Δx and the height of the box, which is $f(x_i)$. So the sum of the areas of the n rectangles is

$$\sum_{i=1}^{n} f(x_i)\,\Delta x$$

Before using Maple to compute this sum, a formula for x_i must be derived. Note the following pattern: $x_1 = a + \Delta x,\quad x_2 = a + 2\Delta x,\quad x_3 = a + 3\Delta x$ and so forth. So in general

$$x_i = a + i\Delta x$$

Now these formulas can be entered into Maple. First, enter f as a function.

> `f:=x->x^2;`

$$f := x \to x^2$$

Then enter Δx (labeled `Dx` in the Maple session) and x_i:

> `Dx:=(b-a)/n; x[i]:=a+i*Dx;`

$$Dx := \frac{b-a}{n}$$

$$x_i := a + \frac{i(b-a)}{n}$$

The above sum is entered with Maple's `Sum` command.

> `riemsum:=Sum(f(x[i])*Dx,i=1..n);`

$$riemsum := \sum_{i=1}^{n} \frac{(a + \frac{i(b-a)}{n})^2 (b-a)}{n}$$

This sum can also be displayed without computing Δx and x_i by using the `rightsum` command from the student package.

> `with(student):`

> `rightsum(f(x),x=a..b,n);`

$$\frac{(b-a)\left(\sum_{i=1}^{n}(a + \frac{i(b-a)}{n})^2\right)}{n}$$

The `Sum` command with an upper case `S` and the `rightsum` command display the sum without calculating its value, as illustrated above. This allows the sum to be checked for typing errors. The `value(%);` command can be added on the same line to evaluate this sum. Since the result is long, we suppress the output of `value(%)`. However it is significantly simplified if we expand the result:

> `riemsum:=Sum(f(x[i])*Dx,i=1..n); value(%):`

$$riemsum := \sum_{i=1}^{n} \frac{(a + \frac{i(b-a)}{n})^2 (b-a)}{n}$$

> `riemsum:=expand(%);`

6.2. THE COMPUTATION OF INTEGRALS

$$riemsum := -\frac{ab^2}{2n} - \frac{a^3}{3} - \frac{a^2b}{2n} + \frac{a^2b}{2n^2} - \frac{ab^2}{2n^2} + \frac{b^3}{3} + \frac{a^3}{2n} - \frac{a^3}{6n^2} + \frac{b^3}{2n} + \frac{b^3}{6n^2}$$

This output represents the sum of the areas of the n rectangles that approximates the area under the graph of $f(x) = x^2$ for $a \leq x \leq b$. To get the precise area under the graph, take the limit of this expression as n approaches infinity.

> `Limit(riemsum,n=infinity); area:=value(%);`

$$\lim_{n \to \infty} -\frac{ab^2}{2n} - \frac{a^3}{3} - \frac{a^2b}{2n} + \frac{a^2b}{2n^2} - \frac{ab^2}{2n^2} + \frac{b^3}{3} + \frac{a^3}{2n} - \frac{a^3}{6n^2} + \frac{b^3}{2n} + \frac{b^3}{6n^2}$$

$$area := -\frac{a^3}{3} + \frac{b^3}{3}$$

This limit of the sum of the areas of rectangles (as the number of rectangles tends to infinity) is the definition of the integral of f from $x = a$ to $x = b$, and is denoted by the symbol

$$\int_a^b f(x)\,dx$$

For a nonnegative function, such as $f(x) = x^2$, the integral measures the area under the curve. However, the definition of the integral holds even for functions which are not nonnegative.

Maple has a command that integrates expressions directly. (Since `f` has been entered as a function, we must evaluate it to get the expression `f(x)`.) To integrate `f(x)`, you enter

> `Int(f(x),x=a..b); area:=value(%);`

$$\int_a^b x^2\,dx$$

$$area := -\frac{a^3}{3} + \frac{b^3}{3}$$

Like the `Sum` and `Diff` commands, the `Int` command with an uppercase I displays the integral without evaluating it, so that it can be checked for typing errors. It is evaluated by adding the `value(%);` command.

Notice that the resulting area is the same as

$$F(b) - F(a) \quad \text{where} \quad F(x) = \frac{x^3}{3}.$$

Further, note that $F(x) = \frac{x^3}{3}$ is an antiderivative of $f(x) = x^2$. This is an illustration of the *Fundamental Theorem of Calculus*, which states that the definite integral $\int_a^b f(x)\,dx$ (defined as the limit of the sum of areas of rectangles) is the same as $F(b) - F(a)$, where F is an antiderivative of f.

As a further illustration of the Fundamental Theorem, try repeating the above procedure with the functions $f(x) = x^3$ and $f(x) = x^4$. The only Maple command that you must change is the one that involves the definition of the function f. The other statements can be re-executed without change.

The antiderivative of f (or indefinite integral) can also be evaluated.

> `Int(f(x),x); value(%);`

$$\int x^2\, dx$$

$$\frac{x^3}{3}$$

Note that Maple *does not insert the constant of integration*.

If f is defined as an expression rather than a function (via `f:=x^2;`) the same syntax as above is used except that `f` is typed instead of `f(x)`. That is, `Int(f,x=a..b);` will display the definite integral $\int_a^b x^2\, dx$, while `Int(f,x);` will display the indefinite integral $\int x^2\, dx$ and then `value(%)` will evaluate them.

Maple cannot find an antiderivative for every expression. For example, try integrating $\sqrt{x^5+1}\, e^x$ with Maple. Some definite integrals have to be approximated either by adding up rectangles or by other more sophisticated techniques discussed later in this chapter. Maple makes it easy to evaluate an approximation to a definite integral by using `evalf`. For example, the integral $\int_1^2 \sqrt{x^5+1}\, e^x\, dx$ can be approximated by entering:

> `Int(sqrt(x^5+1)*exp(x),x=1..2); evalf(%);`

$$\int_1^2 \sqrt{x^5+1}\, e^x\, dx$$

16.38809099

6.3 Integration by Substitution (Change of Variables)

Suppose you want to compute the integral $A = \int \dfrac{1}{\sqrt{4+9x^2}}\, dx$ by making a change of variables (also called a substitution). Actually, Maple can integrate this without changing variables, but it's easier to understand a technique when it is demonstrated on an easy example.

To do this by hand, you first need to identify a relation between the old variable x and a new variable. In this case, take

$$3x = 2\tan\theta$$

where the new variable is θ. Compute the differential and solve for dx. You now have

$$3\, dx = 2\sec^2\theta\, d\theta$$

6.3. INTEGRATION BY SUBSTITUTION (CHANGE OF VARIABLES)

$$dx = \frac{2}{3}\sec^2\theta\, d\theta$$

Next, substitute x and dx into the integral to get

$$A = \int \frac{2\sec^2\theta}{3\sqrt{4+4\tan^2\theta}}\, d\theta = \frac{1}{3}\int \sec\theta\, d\theta = \frac{1}{3}\ln|\sec\theta + \tan\theta| + C$$

Finally, substitute back to get

$$A = \frac{1}{3}\ln\left|\frac{\sqrt{4+9x^2}}{2} + \frac{3x}{2}\right| + C$$

To do this computation using Maple, first enter the integral and the equation relating the old and new variables.

> `A:=Int(1/sqrt(4+9*x^2),x);`

$$A := \int \frac{1}{\sqrt{4+9x^2}}\, dx$$

> `eq:=3*x=2*tan(theta);`

$$eq := 3x = 2\tan(\theta)$$

(Check you typed them correctly.) Then use the `changevar` command in the `student` package, which has three arguments: the first argument is the relation between the old and new variables; the second argument is the integral; and the third argument is the name of the new variable. So load the student package, execute the change of variables, and evaluate the integral (using `value`).

> `with(student):`
> `changevar(eq,A,theta); Atemp:=value(%);`

$$\int \frac{1}{3}\frac{2+2\tan(\theta)^2}{\sqrt{4+4\tan(\theta)^2}}\, d\theta$$

$$Atemp := \frac{1}{3}\operatorname{arcsinh}(\tan(\theta))$$

Finally, to substitute back, solve for the new variable and use the `eval` command.

> `Avalue:=eval(Atemp, theta=arctan(3*x/2));`

$$Avalue := \frac{1}{3}\operatorname{arcsinh}\left(\frac{3x}{2}\right)$$

This is the answer up to an additive constant. Maple's answer looks different from the one we obtained by hand, but the `convert` command shows that they are equivalent.

> `convert(Avalue,ln);`

$$\frac{1}{3}\ln\left(\frac{3x}{2} + \frac{\sqrt{4+9x^2}}{2}\right)$$

(Maple uses the formula arcsinh$(w) = \ln\left(\sqrt{w^2+1}+w\right)$ for the conversion.) However, as with any indefinite integral, check the answer by differentiating the results. Thus,

> `diff(Avalue,x);`

$$\frac{1}{\sqrt{4+9x^2}}$$

gives back the original integrand in A.

If, on applying the `changevar` command, the integral becomes more complicated, go back and try a different substitution or a different integration trick.

You may ask why the `changevar` command is used when the commands

> `A:=Int(1/sqrt(4+9*x^2),x); value(%);`

$$A := \int \frac{1}{\sqrt{4+9x^2}}\,dx$$

$$\frac{1}{3}\operatorname{arcsinh}\left(\frac{3x}{2}\right)$$

will give the same result. The answer is that there are integrals that Maple cannot compute directly. Then an intelligent human-computer interaction may produce a result that Maple could not get on its own. For examples of such integrals, see the exercises.

6.4 Integration by Parts

Suppose you want to compute the integral $\int x \sin x\, dx$ by the method of integration by parts. The first thing to do is to identify u and dv. In this case take

$$u = x \quad \text{and} \quad dv = \sin x\, dx$$

Then compute du and v:

$$du = dx \quad \text{and} \quad v = -\cos x$$

The integration by parts formula $\int u\, dv = uv - \int v\, du$ gives

$$\int x \sin x\, dx = -x\cos x + \int \cos x\, dx = -x\cos x + \sin x + C$$

To do this computation using Maple, first enter the integral.

> `A:=Int(x*sin(x),x);`

$$A := \int x \sin(x)\, dx$$

(Check you typed it correctly.) Then use the `intparts` command in the **student** package, which has two arguments: the first argument is the integral and the second argument is the part of the integrand that will be taken as u. So, load the **student** package, execute the integration by parts, and evaluate the integral (using `value`).

```
>   with(student):
>   intparts(A,x); Aparts:=value(%);
```
$$-x\cos(x) - \int -\cos(x)\,dx$$
$$Aparts := -x\cos(x) + \sin(x)$$

Again, we could have combined these commands as
```
>   value(intparts(Int(x*sin(x),x),x));
```
$$-x\cos(x) + \sin(x)$$
but it is better to work the problem in steps in order to verify that the integral is entered correctly.

If the integral becomes more complicated on applying the `intparts` command, go back and try a different u or a different integration technique. Alternatively, it may be necessary to integrate by parts more than once. Here is an example:
```
>   B:=Int(x^2*sin(x),x);
```
$$B := \int x^2 \sin(x)\,dx$$
```
>   B1:=intparts(B,x^2);
```
$$B1 := -x^2 \cos(x) - \int -2x\cos(x)\,dx$$
```
>   intparts(B1,x); value(%);
```
$$-x^2 \cos(x) + 2x\sin(x) + \int -2\sin(x)\,dx$$
$$-x^2 \cos(x) + 2x\sin(x) + 2\cos(x)$$

6.5 Integration using Partial Fractions

Maple computes partial fractions by using the `convert` command with the `parfrac` option. This requires three arguments: a rational function, the keyword `parfrac` and the independent variable.

For example, to evaluate $\int \dfrac{x^2 - 3x + 1}{x^3 + x^2 - 2x}\,dx$, define the integrand
```
>   f:=(x^2-3*x+1)/(x^3+x^2-2*x);
```
$$f := \frac{x^2 - 3x + 1}{x^3 + x^2 - 2x}$$
Find its partial fraction expansion
```
>   fpar:=convert(f,parfrac,x);
```
$$fpar := \frac{11}{6(x+2)} - \frac{1}{3(x-1)} - \frac{1}{2x}$$
Construct the integral and find its value

```
>  Int(fpar,x); value(%);
```
$$\int \frac{11}{6(x+2)} - \frac{1}{3(x-1)} - \frac{1}{2x}\,dx$$
$$\frac{11}{6}\ln(x+2) - \frac{1}{3}\ln(x-1) - \frac{1}{2}\ln(x)$$

Note that Maple omits the *absolute values* in its answer. It simply gives *an* antiderivative!

Of course, Maple could compute this directly
```
>  Int(f,x); value(%);
```
$$\int \frac{x^2 - 3x + 1}{x^3 + x^2 - 2x}\,dx$$
$$\frac{11}{6}\ln(x+2) - \frac{1}{3}\ln(x-1) - \frac{1}{2}\ln(x)$$

but it may be useful to be able to check intermediate results in your hand computations using partial fractions.

6.6 Approximate Integration

When you can't compute an integral exactly, you may need to approximate it numerically using a left Riemann sum, a right Riemann sum, a midpoint Riemann sum, the trapezoid rule or Simpson's rule. At the beginning of this chapter, we discussed Maple's commands for left and right Riemann sums. In this section, we also discuss Maple's commands for midpoint Riemann sums, the trapezoid rule and Simpson's rule. We start by loading the `student` package
```
>  with(student):
```
which contains the Maple commands `leftsum`, `rightsum`, `middlesum`, `trapezoid` and `simpson`. Let's use these commands to approximate

$$\int_0^1 \sin(x^2+1)\,dx$$

First define the function
```
>  f:=x->sin(x^2+1);
```
$$f := x \to \sin(x^2 + 1)$$

The left, right and midpoint Riemann sums, trapezoid rule and Simpson's rule with 10 intervals are computed as follows:
```
>  leftsum(f(x),x=0..Pi,10); l:=evalf(%);
```
$$\frac{1}{10}\pi \left(\sum_{i=0}^{9} \sin\left(\frac{i^2\pi^2}{100} + 1\right) \right)$$
$$l := 1.173839061$$
```
>  rightsum(f(x),x=0..Pi,10); r:=evalf(%);
```

6.6. APPROXIMATE INTEGRATION

$$\frac{1}{10}\pi\left(\sum_{i=1}^{10}\sin(\frac{i^2\pi^2}{100}+1)\right)$$

$$r := 0.5978131996$$

> `middlesum(f(x),x=0..Pi,10); m:=evalf(%);`

$$\frac{1}{10}\pi\left(\sum_{i=0}^{9}\sin\left(\frac{(i+\frac{1}{2})^2\pi^2}{100}+1\right)\right)$$

$$m := 0.8977656549$$

> `trapezoid(f(x),x=0..Pi,10); t:=evalf(%);`

$$\frac{\pi}{30}\left(\sin(1)+\sin(\pi^2+1)+4\left(\sum_{i=1}^{5}\sin(\frac{(2i-1)^2\pi^2}{100}+1)\right)+2\left(\sum_{i=1}^{4}\sin(\frac{i^2\pi^2}{25}+1)\right)\right)$$

$$t := 0.8858261305$$

> `simpson(f(x),x=0..Pi,10); s:=evalf(%);`

$$\frac{1}{30}\pi\left(\sin(1)+\sin(\pi^2+1)+4\left(\sum_{i=1}^{5}\sin(\frac{(2i-1)^2\pi^2}{100}+1)\right)+2\left(\sum_{i=1}^{4}\sin(\frac{i^2\pi^2}{25}+1)\right)\right)$$

$$s := 0.9058553279$$

The last three are fairly close to Maple's result

> `Int(f(x),x=0..Pi); i:=evalf(%);`

$$\int_0^\pi \sin(x^2+1)\,dx$$

$$i := 0.8934801800$$

The error using Simpson's rule is

> `E[S]:=s-i;`

$$E_S := 0.0123751479$$

Is this within the bound determined by the Simpson's rule error formula? Recall this formula says

$$|E_S| \leq \frac{K(b-a)^5}{180n^4} \qquad \text{where} \quad K \geq \left|f^{(4)}(x)\right| \quad \text{for all} \quad a \leq x \leq b$$

Here:

> `n:=10; (a,b):=(0,Pi);`

$$n := 10$$
$$a, b := 0, \pi$$

To find K, we compute $f^{(4)}(x)$ (You wouldn't want to do this by hand!) and plot its absolute value:

> `d4f:=(D@@4)(f);`

$$d4f := x \to 16\sin(x^2+1)x^4 - 48\cos(x^2+1)x^2 - 12\sin(x^2+1)$$

```
> plot(abs(d4f(x)),x=0..Pi);
```

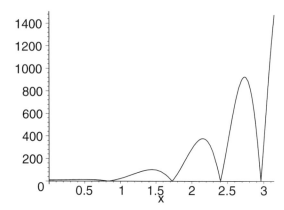

The maximum occurs at $x = \pi$. So we define K as

```
> K:=evalf(abs(d4f(Pi)));
```
$$K := 1474.771668$$

So the error from Simpson's rule is bounded by

```
> error_bound:=evalf(K*(b-a)^5/(180*n^4));
```
$$error\_bound := 0.2507273118$$

Yes, the error we found $E_S = .0123751482$ is less than this error bound.

OK, now let's use the error bound for Simpson's rule to determine a value of n for which Simpson's rule approximates the integral within 10^{-5}. We unassign n and solve the inequality $E_S < 10^{-5}$.

```
> n:='n':
> eq:=evalf(K*(b-a)^5/(180*n^4)=10^(-5));
```
$$eq := \frac{2507.273118}{n^4} = 0.00001000000000$$

```
> fsolve(eq,n=0..infinity);
```
$$125.8346980$$

Thus n must be 126 or larger. Consequently, if we apply Simpson's rule with $n = 126$ then the result should have an error less than 10^{-5}.

```
> s:=evalf(simpson(f(x),x=0..Pi,126));
```
$$s := 0.8934803287$$

```
> E[S]:=s-i;
```
$$E_S := 0.1487\,10^{-6}$$

It does!

Remark: Maple uses a much more sophisticated method for numerically approximating integrals but the midpoint rule or Simpson's rule is usually sufficient.

6.7 Summary

- Load the student package and use `leftbox` and `rightbox` to help visualize rectangles in Riemann sums. Here are the steps:

`with(student);`	Load the student package.
`f:=x-> expr;`	Define the function.
`leftbox(f(x), x=a..b,n);`	Plot the left Riemann sum.
`rightbox(f(x), x=a..b,n);`	Plot the right Riemann sum.

- Use Maple to set up a Riemann sum over the interval from a to b using n rectangles. Compute the area as the limit of the Riemann sums. Here are the steps:

`f:=x-> expr;`	Define the function.
`Dx:=(b-a)/n;`	Compute the rectangle width, with a specific number `n`.
`Sum(f(a+i*Dx)*Dx, i=1..n);` `area:=value(%);`	Compute the Riemann sum.
`Limit(area,n=infinity);` `value(%);`	Compute the limit, which is the area.

- Set up Riemann sums with `leftsum` and `rightsum` which are in the `student` package.

`leftsum(f(x), x=a..b,n);`	Set up a left Riemann sum.
`rightsum(f(x), x=a..b,n);`	Set up a right Riemann sum.

- Compute indefinite and definite integrals using `Int` and `value`

`Int(f(x),x); value(%);`	Compute an indefinite integral.
`Int(f(x),x=a..b); value(%);`	Compute a definite integral.

- Recall that *Maple does not add a constant of integration.*

- Maple will only integrate an expression. If f is a Maple function, it must be evaluated to $f(x)$ before integrating.

- The Maple commands `changevar`, `intparts` and `convert/parfrac` can be used to perform integration by substitution, integration by parts and integration using partial fractions. Note `changevar` and `intparts` are in the `student` package. These commands can be used to help Maple calculate complicated integrals in closed form.

- Approximate definite integrals using `leftsum`, `rightsum`, `middlesum`, `trapezoid` and `simpson` which are in the `student` package. Perform the sums using `value(%)`.

`leftsum(f(x), x=a..b,n);`	Set up a left Riemann sum.
`rightsum(f(x), x=a..b,n);`	Set up a right Riemann sum.
`middlesum(f(x), x=a..b,n);`	Set up a midpoint Riemann sum.
`trapezoid(f(x), x=a..b,n);`	Set up a trapezoid rule sum.
`simpson(f(x), x=a..b,n);`	Set up a Simpson's rule sum.

6.8 Exercises

1. Load the `student` package via `with(student);`

   (a) Use the `leftbox` command with 9 rectangles to display the left Riemann sum approximation to $\int_0^2 e^{x^2/2}\,dx$ Does this set of rectangles overestimate or underestimate the integral?

   (b) Set up and evaluate the left Riemann sums to approximate the integral with $n = 9, 27, 81$ and 243 rectangles using the `leftsum` command.

   (c) Compute the integral using Maple's `Int` and `evalf(%)` commands as done in the text.

   (d) Compare the values of the Riemann sums in (b) to the value of the integral in (c). (Subtract the exact value from the estimates.) As the number of rectangles triples, what happens to the difference between the value of the Riemann sum and the value of the integral?

2. Repeat Exercise 1 for right Riemann sums, using the `rightbox` and `rightsum` commands.

3. Repeat Exercise 1 for midpoint Riemann sums, using the `middlebox` and `middlesum` commands.

4. Repeat Exercises 1, 2 and 3 with the function $f(x) = \dfrac{1 - x^4}{x^2 + 1}$ using the interval $[0, 1]$.

5. Use Maple to evaluate the following integrals. Check (b) and (c) by differentiating.

   (a) $\int_1^3 x^4\sqrt{x^2+1}\,dx$

   (b) $\int \sin^2(x)\,dx$

   (c) $\int x\cos^4(x)\,dx$

6. There are two ways to evaluate the integral $\int \sec^4(x)\tan(x)\,dx$ by u-substitution. The first is to let $u = \sec(x)$ and the second is to let $u = \tan(x)$. Based on Maple's answer to this integral, (using `Int` and `value`) which substitution is it using? For each substitution, use `changevar`, and `value` and use `eval` to substitute back. These two methods of substitution lead to apparently different answers. Are they, in fact, the same? If they are different, then how do you explain the fact that they are answers to the same problem? (Subtract the answers and `simplify`.)

6.8. EXERCISES

7. Express each of the following limits as an integral and then use Maple to evaluate. (The first two integrals are provided for you.)

   (a) $\lim_{n\to\infty} \sum_{i=1}^{n} \frac{1}{n}\sqrt{4 + \frac{i^2}{n^2}} = \int_0^1 \sqrt{4+x^2}\,dx$

   (b) $\lim_{n\to\infty} \sum_{i=1}^{n} \frac{3}{n}\sqrt{9 + \frac{9i^2}{n^2}} = \int_0^3 \sqrt{9+x^2}\,dx$

   (c) $\lim_{n\to\infty} \sum_{i=1}^{n} \frac{\pi}{n} \cos^4\left(\frac{i\pi}{n}\right)$

   (d) $\lim_{n\to\infty} \sum_{i=1}^{n} \frac{3}{n}\left(\left(2+\frac{3i}{n}\right)^4 + 6\left(2+\frac{3i}{n}\right)\right)$

8. This problem is a continuation of Exercise 9 in Chapter 5. In that exercise, the northern boundary of a plot of land is described by the following table (each unit represents 100 feet).

x-value	0.0	1.0	2.0	3.0	4.0
y-value	3.2	1.5	2.6	3.2	2.7

   where the x-axis represents the (straight) southern boundary with the origin located at its western corner. (The eastern and western boundaries are also straight.) Use these data to compute an approximation to the area of this plot of land using trapezoids.

9. Now do Exercise 9 from Chapter 5 or load the answer from a file if you have already done that exercise. Use the quadratic and cubic functions described therein to evaluate the area of this plot of land.

10. *Area of Texas.* As in Exercise 2 in Chapter 3, the northern and southern boundaries of the state of Texas are given by the following data.
    north:=[[0,0],[3,0],[3,4.5],[6,4.5],[6,2.2],[7,2.1],[8,1.8],
    [9,1.9],[10,1.8],[11,1.7],[11,-2.2]];
    south:=[[0,0],[1,-1.1],[2,-2.5],[3,-2.9],[4,-2.3],[5,-2.8],
    [6,-4.4],[7,-5.8],[8,-6.1],[9,-3.3],[10,-2.8],[11,-2.2]];
    (Enter these in Maple.) Here, the origin is the westernmost corner of Texas (near El Paso) and the x-axis is the extension of the east-west border between New Mexico and Texas. Each unit represents approximately 69 miles. Use these data to approximate the area of Texas by computing a left Riemann sum formed from rectangles whose widths are 1 unit and whose heights are determined by the second coordinates of the data.
    HINT: To refer to the entries in a list, use square brackets [ ]. For example, south[3] refers to the point $[2,-2.5]$ and south[3][2] refers to the second entry of this point, -2.5. So to (symbolically) sum up all second entries of the points on this list, enter Sum(south[i][2],i=1..11); Then, to actually compute the sum, enter value(%);

11. *Area of Texas, Two.* Improve your estimate from Exercise 10 by using trapezoids instead of rectangles.

12. *Area of Texas, Three.* Improve your estimate from Exercise 10 by using the quadratic and cubic curves found in Exercise 10 of Chapter 5.

13. Evaluate the integral $\int \dfrac{x^5 \arctan(x^2)}{(1+x^4)^3}\,dx$. First try the `value` command by itself. The answer is not useful because it is given in terms of complex numbers. Then use the `changevar` command with $u = \arctan(x^2)$, followed by `value`, `eval`, and `simplify`. Be sure to check your answer using `diff`, and `simplify`.

14. Evaluate the integral $\int x^2 \ln(x + \sqrt{1+x^2})\,dx$. First try the `value` command by itself. Maple gives up. Then use the `intparts` and `value` commands with $u = x^2$. Can Maple do it now? Finally, use the `intparts` and `value` commands with $u = \ln(x + \sqrt{1+x^2})$. Can Maple do it now? Be sure to check the answer.

15. Find the partial fraction expansion for $f = \dfrac{x^6 + 3x^2 - 1}{(x+2)^2(x^2+4)^2}$ and use it to evaluate $\int \dfrac{x^6 + 3x^2 - 1}{(x+2)^2(x^2+4)^2}\,dx$.

16. Compute $\int \dfrac{x^3}{x^8 + 1}\,dx$

17. Compute $\int \dfrac{\sin(\sqrt{x})}{2\sqrt{x}}\,dx$

18. Compute $\int \cos^2(x) \sin^4(x)\,dx$

19. Compute $\int_{-1}^{3} \dfrac{x^5 + x}{\sqrt{1+x^2}}\,dx$

20. Compute $\int_{\pi/4}^{3\pi/4} \sqrt{\csc^2(x) - 1}\,dx$

21. Repeat Exercise 1 b,c,d using the trapezoid rule instead of left Riemann sums for the integral $\int_0^2 e^{x^2/2}\,dx$.

    Explain carefully why we cannot repeat the exercise using Simpson's rule.

22. Use the error estimate for Simpson's rule to determine a value for n so that if Simpson's rule is used to approximate the integral $\int_0^2 e^{x^2/2}\,dx$, then the value is accurate to 15 decimal places. Calculate the Simpson's rule approximation for this n and check its accuracy.

Chapter 7

Applications of Integration

The first three sections in this chapter use Maple to compute the standard geometric applications of integration: areas, volumes, arc lengths and surface areas. The last section uses Fourier series to approximate functions by sums of multiples of sines and cosines.

7.1 Area

We present two examples in this section: the area between a curve and the x-axis, and the area between two curves.

EXAMPLE 1: Find the area that lies between the graph of

$$f(x) = -0.128x^3 + 1.728x^2 - 5.376x + 2.864$$

and the x-axis.

SOLUTION: Start by inputting an expression for f into Maple.

```
>   f:=-0.128*x^3+1.728*x^2-5.376*x+2.864;
```
$$f := -0.128\,x^3 + 1.728\,x^2 - 5.376\,x + 2.864$$

Now plot f over an interval that shows all points where f crosses the x-axis. By trial and error, the interval $-2 \leq x \leq 10$ will do.

```
>   plot(f,x=-2..10);
```

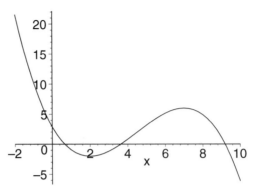

The graph of f crosses the x-axis at three points—between 0 and 1, between 3 and 4, and between 9 and 10. Use `fsolve` to find these roots and assign them to variablea, say a, b and c.

> `a,b,c:=fsolve(f=0,x);`

$$a, b, c := 0.6697777844, 3.631759112, 9.198463104$$

Note that all the roots can be found with `fsolve` without specifying a range for x, because f is a polynomial. From the plot, it is evident that the graph of f is below the x-axis between a and b and above the x-axis between b and c. Therefore, the area is the following sum of integrals:

$$A = -\int_a^b f\,dx + \int_b^c f\,dx$$

In Maple we enter

> `-Int(f,x=a..b)+Int(f,x=b..c); A:=value(%);`

$$-\int_{0.6697777844}^{3.631759112} -0.128\,x^3 + 1.728\,x^2 - 5.376\,x + 2.864\,dx$$
$$+ \int_{3.631759112}^{9.198463104} -0.128\,x^3 + 1.728\,x^2 - 5.376\,x + 2.864\,dx$$

$$A := 25.05016797$$

As mentioned in the last chapter, the `Int` command (with an uppercase I) displays the integral without computing its value so that it can be checked for typing errors. Then the `value` command evaluates it. The area of interest is about 25.05 square units.

EXAMPLE 2: Compute the area between the graph of f (defined above) and the graph of

$$g(x) = \ln x$$

SOLUTION: Once again, we start by defining g as an expression in Maple.

> `g:=ln(x);`

$$g := \ln(x)$$

Now plot f and g on the same coordinate axes with the command
```
> plot({f,g},x=-2..10);
```

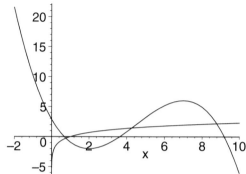

The plot shows three points of intersection, which can again be found with `fsolve`, but this time you need to specify ranges since g is not a polynomial.
```
> a:=fsolve(f=g,x); b:=fsolve(f=g,x=4..5);
> c:=fsolve(f=g,x=8..9);
```
$$a := 0.7587367455$$
$$b := 4.270958448$$
$$c := 8.793410794$$

The graph of f is below the graph of g between **a** and **b** while g is below f between **b** and **c**. Therefore, the area between the graphs is

$$A = \int_a^b (g-f)\,dx + \int_b^c (f-g)\,dx$$

In Maple this is entered as
```
> Int(g-f,x=a..b)+Int(f-g,x=b..c); A:=value(%);
```

$$\int_{0.7587367455}^{4.270958448} \ln(x) + 0.128\,x^3 - 1.728\,x^2 + 5.376\,x - 2.864\,dx$$
$$+ \int_{4.270958448}^{8.793410794} -0.128\,x^3 + 1.728\,x^2 - 5.376\,x + 2.864 - \ln(x)\,dx$$

$$A := 18.17292990$$

7.2 Volume

We look at volume by slicing and volume of revolution.

Volume by Slicing: To calculate a volume by slicing, the first step is to derive an expression that represents the cross sectional area. Then this expression is integrated over the appropriate interval.

EXAMPLE 1: The base of a solid is bounded by the curves $y = -x^2 + 5x - 2$ and $y = x$. The cross sections perpendicular to the x-axis are equilateral triangles. Find the volume of the solid.

SOLUTION: First, define f and g in Maple and graph them

```
> f:=-x^2+5*x-2; g:=x;
```
$$f := -x^2 + 5x - 2$$
$$g := x$$

```
> plot({f,g},x=0..4);
```

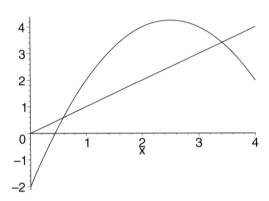

From the plot, notice that f and g cross at two points: one between $x = 0$ and $x = 1$ and the other between $x = 3$ and $x = 4$. Find these roots using fsolve and assign them to the variables a and b

```
> a,b:=fsolve(f=g,x);
```
$$a, b := 0.5857864376, 3.414213562$$

The cross section of this solid is an equilaterial triangle of side

```
> s:=f-g;
```
$$s := -x^2 + 4x - 2$$

The area of an equilaterial triangle is $A = \dfrac{\sqrt{3}\, s^2}{4}$. So the cross sectional area is

```
> A:=sqrt(3)/4*s^2;
```
$$A := \frac{\sqrt{3}\,(-x^2 + 4x - 2)^2}{4}$$

The total volume of this solid is obtained by integrating this expression over the interval $a \leq x \leq b$.

```
> Int(A,x=a..b); V:=value(%);
```
$$\int_{0.5857864376}^{3.414213562} \frac{\sqrt{3}\,(-x^2 + 4x - 2)^2}{4}\, dx$$
$$V := 2.612789059$$

7.2. VOLUME

Volume of Revolution: To calculate a volume of revolution, the first step is to determine if it is an x- or y-integral. The second step is to determine if a slice rotates into a disk, washer or cylindrical shell. The third step is to compute the integral with the appropriate formula. Assuming x-integrals, the formulas are:

$$\int_a^b \pi R^2 \, dx \quad \ldots\ldots\ldots \quad \text{disk}$$

R is the radius of the disk.

$$\int_a^b \pi (R^2 - r^2) \, dx \quad \ldots \quad \text{washer}$$

R and r are the outer and inner radii of the washer.

$$\int_a^b 2\pi r h \, dx \quad \ldots\ldots\ldots \quad \text{cylinder (shells)}$$

r and h are the radius and height of the cylinder.

EXAMPLE 2: Consider the region that is bounded by the curves $y = -x^2 + 5x - 2$ and $y = x$. Find the volume of the solid that is obtained by revolving this region about the x-axis.

SOLUTION: The region is the same as in Example 1. As in that example, we want to do an x-integral. So we graph the curves with a slice perpendicular to the x-axis.

```
> plot([f,g, [[2.5,subs(x=2.5,f)], [2.5,subs(x=2.5,g)]]],
>   x=0..4, thickness = [2,2,5]);
```

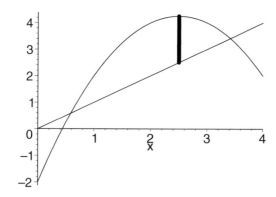

Notice that the third quantity plotted is the line segment from $(2.5, f(2.5))$ to $(2.5, g(2.5))$ and it is given thickness 5 by putting the quantities to be plotted and the corresponding parameters in square brackets. When this slice is rotated about the x-axis, it sweeps out a washer. The outer radius is f and the inner radius is g. So the volume is

```
> Int(Pi*(f^2-g^2),x=a..b); V:=value(%);
```

$$\int_{0.5857864376}^{3.414213562} \pi\left(\left(-x^2 + 5x - 2\right)^2 - x^2\right) dx$$

$$V := 66.34705188$$

EXAMPLE 3: Consider the region that is bounded by the curves $y = -x^2+5x-2$ and $y = x$. Find the volume of the solid that is obtained by revolving this region about the y-axis.

SOLUTION: The region is the same as in Examples 1 and 2. Again we want to do an x-integral. The graph of the curves and a slice perpendicular to the x-axis is the same as in Example 2, but when this slice is rotated about the y-axis, it sweeps out a cylindrical shell. The radius is x and the height is f-g. So the volume is

> `Int(2*Pi*x*(f-g),x=a..b); V:=value(%);`

$$\int_{0.5857864376}^{3.414213562} 2\pi x \left(-x^2 + 4x - 2\right) dx$$

$$V := 47.39075134$$

7.3 Arc Length and Surface Area

Most textbook problems on arc length and surface area are contrived so that the integrals are doable. Not so in the real world.

Arc Length of a Lissajous Figure: A Lissajous figure is a curve in the plane which may be parametrized as

$$(x, y) = (\cos(pt), \sin(qt))$$

where p and q are positive integers. For example here is a Lissajous figure with $p = 3$ and $q = 4$.

> `plot([cos(3*t),sin(4*t),t=0..2*Pi]);`

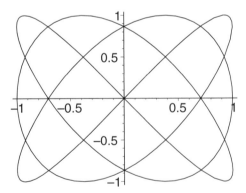

As t varies from 0 to 2π, the x-coordinate oscillates $p = 3$ times while the y-coordinate oscillates $q = 4$ times. (Here, 3 bumps on the left and right and 4 bumps on the top and bottom.)

7.3. ARC LENGTH AND SURFACE AREA

EXAMPLE 1: Find the arc length of the Lissajous figure with $p = 3$ and $q = 4$.
SOLUTION: The arc length of a parametric curve is given by

$$L = \int_a^b \sqrt{\left(\frac{dx}{dt}\right)^2 + \left(\frac{dy}{dt}\right)^2} \, dt$$

So define $x(t)$ and $y(t)$ and take their derivatives:

```
> xt,yt:=cos(3*t),sin(4*t);
```
$$xt, yt := \cos(3t), \sin(4t)$$

```
> Dxt:=diff(xt,t); Dyt:=diff(yt,t);
```
$$Dxt := -3\sin(3t)$$
$$Dyt := 4\cos(4t)$$

Then the arclength is

```
> Int(sqrt(Dxt^2+Dyt^2),t=0..2*Pi); L:=value(%);
```

$$\int_0^{2\pi} \sqrt{9\sin(3t)^2 + 16\cos(4t)^2} \, dt$$

$$L := \int_0^{2\pi} \sqrt{9\sin(3t)^2 + 16\cos(4t)^2} \, dt$$

Oops! Maple cannot do it exactly. (No one can!) So do it numerically:

```
> L:=evalf(%);
```
$$L := 21.23721523$$

Surface Area of Revolution: When a curve $y = f(x)$ is rotated about an axis, the area of the resulting surface is given by

$$A = \int_a^b 2\pi r \sqrt{1 + \left(\frac{dy}{dx}\right)^2} \, dx$$

where r is the radius of revolution. Specifically, $r = x$ if it is rotated about the y-axis and $r = y = f(x)$ if it is rotated about the x-axis.

EXAMPLE 2: The curve $y = 3 + \cos x$ for $0 \le x \le 4\pi$ is revolved about the x-axis. Find the surface area.

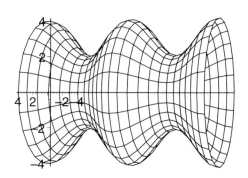

SOLUTION: Enter the function and compute its derivative:
```
> y1:=3+cos(x);
```
$$y1 := 3 + \cos(x)$$
```
> Dy1:=diff(y1,x);
```
$$Dy1 := -\sin(x)$$
The curve is rotated about the x-axis; so $r = y$. So the surface area is
```
> Int(2*Pi*y1*sqrt(1+Dy1^2),x=0..4*Pi); A:=value(%);
```
$$\int_0^{4\pi} 2\pi (3 + \cos(x)) \sqrt{1 + \sin(x)^2}\, dx$$

$$A := 48\pi\sqrt{2}\,\text{EllipticE}(\frac{\sqrt{2}}{2}) - \pi \ln(\sqrt{2}-1) + \frac{1}{2}\pi \ln(3 - 2\sqrt{2})$$

The EllipticE function is not very informative. So get a decimal approximation:
```
> A:=evalf(%);
```
$$A := 288.0361274$$

EXAMPLE 3: The curve $y = 3 + \cos x$ for $0 \le x \le 4\pi$ is revolved about the y-axis. Find the surface area.

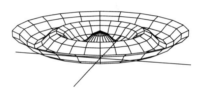

SOLUTION: The curve is the same as in Exercise 2, but it is rotated about the y-axis; so $r = x$. So the surface area is
```
> Int(2*Pi*x*sqrt(1+Dy1^2),x=0..4*Pi); A:=value(%);
```
$$\int_0^{4\pi} 2\pi x \sqrt{1 + \sin(x)^2}\, dx$$

$$A := \int_0^{4\pi} 2\pi x \sqrt{1 + \sin(x)^2}\, dx$$

Maple cannot do it exactly. So do it numerically:
```
> A:=evalf(%);
```
$$A := 603.2614546$$

7.4 Introduction to Fourier Series - Cosine Expansions

This is an optional section.

One of the interesting applications of integration is the approximation of functions. We will demonstrate in this section that smooth functions can be

7.4. INTRODUCTION TO FOURIER SERIES - COSINE EXPANSIONS

approximated by sums of trigonometric functions called Fourier series. More precisely, we will demonstrate that a function on an interval of the form $[0, L]$ can be approximated by sums of the functions

$$1, \cos\left(\frac{\pi x}{L}\right), \cos\left(\frac{2\pi x}{L}\right), \ldots, \cos\left(\frac{n\pi x}{L}\right), \ldots \tag{7.1}$$

To do this, we need to learn some properties of the functions above. First, notice that

$$\int_0^L 1 \, dx = L.$$

and if k is a positive integer, then

$$\int_0^L \cos\left(\frac{k\pi x}{L}\right) dx = \frac{L}{k\pi} \sin(k\pi) = 0$$

Further, if k is a positive integer, then

$$\int_0^L \cos^2\left(\frac{k\pi x}{L}\right) dx = \frac{L}{2}\left(1 + \frac{\sin 2k\pi}{2k\pi}\right) = \frac{L}{2}$$

and if k and m are positive integers with $k \neq m$, then

$$\int_0^L \cos\left(\frac{k\pi x}{L}\right) \cos\left(\frac{m\pi x}{L}\right) dx = 0$$

(Do the integration and see for yourself!)

Let's use this information to find a way to approximate a function $f(x)$ in the form

$$f(x) \approx a_0 + a_1 \cos\left(\frac{\pi x}{L}\right) + \cdots + a_n \cos\left(\frac{n\pi x}{L}\right) \tag{7.2}$$

on the interval $[0, L]$.

In order to determine the coefficient a_0, we simply integrate both sides of (7.2) above and obtain

$$\int_0^L f(x) \, dx = a_0 L$$

(since all the other terms integrate to zero). This gives

$$a_0 = \frac{1}{L} \int_0^L f(x) \, dx \tag{7.3}$$

(Note that a_0 is the average value of $f(x)$ on the interval $[0, L]$.)

Next, we obtain a formula for a_1 by multiplying both sides of (7.2) by $\cos\left(\frac{\pi x}{L}\right)$ and integrating. This yields

$$\int_0^L f(x) \cos\left(\frac{\pi x}{L}\right) dx = \int_0^L a_1 \cos^2\left(\frac{\pi x}{L}\right) dx = a_1 \frac{L}{2}$$

since all the other terms integrate to zero. Consequently, we have

$$a_1 = \frac{2}{L} \int_0^L f(x) \cos\left(\frac{\pi x}{L}\right) dx$$

a_2 can be found in a similar manner by multiplying both sides of (7.2) by $\cos\left(\frac{2\pi x}{L}\right)$ and integrating. In this case we obtain

$$a_2 = \frac{2}{L} \int_0^L f(x) \cos\left(\frac{2\pi x}{L}\right) dx$$

In general, for $k = 1, 2, \ldots, n$, we obtain

$$a_k = \frac{2}{L} \int_0^L f(x) \cos\left(\frac{k\pi x}{L}\right) dx \tag{7.4}$$

The numbers $a_0, a_1, \ldots, a_n$ are called the **Fourier cosine coefficients** of the function $f(x)$ on the interval $[0, L]$.

DEFINITION: Let $f : [0, L] \to \mathbb{R}$ be a smooth function. If n is a positive integer then the nth order **Fourier cosine expansion** for $f(x)$ on the interval $[0, L]$ is

$$a_0 + \sum_{k=1}^n a_k \cos\left(\frac{k\pi x}{L}\right)$$

where a_0 and a_k are given by the formulas in (7.3) and (7.4) respectively.

EXAMPLE: Compute the 8th order Fourier cosine expansion for the function $f(x) = x^2 - 1$ on the interval $[0, 3]$. Plot $f(x)$ and its Fourier cosine expansion.

SOLUTION: Using the formulas above, we can input f and compute the Fourier coefficients as follows:

> f:=x^2-1;

$$f := x^2 - 1$$

> 1/3*Int(f,x=0..3); a[0]:=value(%);

$$\frac{1}{3} \int_0^3 x^2 - 1 \, dx$$

$$a_0 := 2$$

> 2/3*Int(f*cos(k*Pi*x/3),x=0..3);
> a[k]:=value(%) assuming k::posint;

$$\frac{2}{3} \int_0^3 (x^2 - 1) \cos\left(\frac{k\pi x}{3}\right) dx$$

$$a_k := \frac{36(-1)^k}{k^2 \pi^2}$$

Notice the use of the assume facility to say that k is a positive integer. Try this command without the assumption. (Read the help pages by typing ?assume and ?assuming.)

7.4. INTRODUCTION TO FOURIER SERIES - COSINE EXPANSIONS

Consequently, the 8th order Fourier cosine expansion for $f(x) = x^2 - 1$ on the interval $[0, 3]$ is

```
>  fcos:=a[0] + sum(a[k]*cos(k*Pi*x/3),k=1..8);
```

$$fcos := 2 - \frac{36\cos(\frac{\pi x}{3})}{\pi^2} + \frac{9\cos(\frac{2\pi x}{3})}{\pi^2} - \frac{4\cos(\pi x)}{\pi^2} + \frac{9}{4}\frac{\cos(\frac{4\pi x}{3})}{\pi^2}$$
$$- \frac{36}{25}\frac{\cos(\frac{5\pi x}{3})}{\pi^2} + \frac{\cos(2\pi x)}{\pi^2} - \frac{36}{49}\frac{\cos(\frac{7\pi x}{3})}{\pi^2} + \frac{9}{16}\frac{\cos(\frac{8\pi x}{3})}{\pi^2}$$

Finally, plot $f(x)$ and this expression on $[0, 3]$ to see the quality of the approximation.

```
>  plot({f,fcos},x=0..3);
```

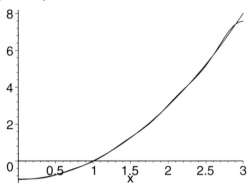

It is almost impossible to see a difference between the curve and the approximation. Besides some wiggles, the only deviation is near the endpoint $x = 3$. The reason for the poor behavior there is simple. All of the functions in (7.1) (with $L = 3$) have a zero derivative at $x = 3$. Since the function $f(x) = x^2 - 1$ does not have this property, we can expect the approximating Fourier expansion to have some trouble near $x = 3$. Notice that there is no a problem at $x = 0$ since $f'(0) = 0$, and all of the functions in (7.1) have this property.

What happens when we look at these graphs on a larger interval? First let's try the interval $[-3, 3]$:

```
>  plot({f,fcos},x=-3..3);
```

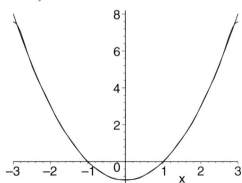

The agreement is quite good. If you are wondering whether this will always happen, then the answer is a definite NO! The reason it happens here is because $f(x) = x^2 - 1$ is an even function, and all of the functions in (7.1) are also even functions!

Finally, let's look at a larger interval, say $[-4, 10]$.

```
> plot({f,fcos},x=-4..10,-5..20);
```

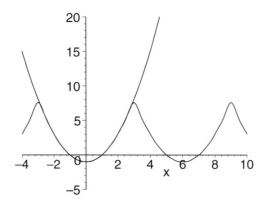

Notice that the Fourier cosine expansion falls away quickly outside the interval $[-3, 3]$. You could say that the expansion did what it was asked to do. However, in actuality, the fall off takes place because all of the functions in (7.1) are periodic with period 6 (since $L = 3$) and $f(x)$ is not a periodic function.

7.5 Summary

- Use `fsolve` to find successive points of intersection of the graphs of Maple expressions f and g. Use `Int` and `value` or `Int` and `evalf` syntax to compute integrals.

- **Area:** Integrate the difference $f - g$ or $g - f$ between intersection points to find the area.

- **Volumes by Slicing:** A slice of the base has length given by the difference $f - g$. Use this to find the area of a cross sectional slice of the solid. Integrate this cross sectional area to get the volume.

$$\int_a^b A(x)\, dx$$

- **Volumes of Revolution:** Determine if you need an x- or y-integral. Determine if a slice rotates into a disk, washer or cylindrical shell. Integrate to get the volume:

$$\int_a^b \pi R^2 \, dx \quad \ldots\ldots\ \text{disk}$$

R is the radius of the disk.

$$\int_a^b \pi(R^2 - r^2) \, dx \ . \quad \text{washer}$$

R and r are the outer and inner radii of the washer.

$$\int_a^b 2\pi r h \, dx \quad \ldots\ldots\ \text{cylinder (shells)}$$

r and h are the radius and height of the cylinder.

- **Arc Length:** Determine if you need an x-, y- or t-integral. Integrate the arc length differential

$$\begin{aligned}
ds &= \sqrt{1 + \left(\frac{dy}{dx}\right)^2}\, dx \\
&= \sqrt{\left(\frac{dx}{dy}\right)^2 + 1}\, dy \\
&= \sqrt{\left(\frac{dx}{dt}\right)^2 + \left(\frac{dy}{dt}\right)^2}\, dt
\end{aligned}$$

to get the arc length.

- **Surface Area:** Determine if you need an x-, y- or t-integral. The surface area is the integral

$$\int_a^b 2\pi r \, ds$$

where ds is the arc length differential and r is the radius of revolution. Specifically, $r = x$ if it is rotated about the y-axis and $r = y$ if it is rotated about the x-axis.

- **Fourier Cosine Series:** Use Int and value to find the Fourier coefficients. Use sum or Sum and value to create the Fourier cosine series approximations.

7.6 Exercises

1. Find the area of the region that is bounded above by the curves $y = 10\ln(x)$ and $y = 4 - x^4 - x$ and below by the x-axis.

2. Consider the region in the previous exercise.

    (a) Find the volume of the solid obtained by revolving this region about the x-axis.

    (b) Find the volume of the solid obtained by revolving this region about the y-axis.

(c) Find the volume of the solid obtained by revolving this region about the line $x = 4$.

(d) Find the volume of the solid obtained by revolving this region about the line $y = 4$.

3. The region in the first quadrant bounded by the coordinate axes and the graph of $y = 15 - 2x^2$ is revolved about the x-axis to form a solid. Use `leftsum` and `evalf` to approximate the volume of this solid by summing the volumes of 20, 40, 200, and 500 disks of equal thickness. Is this an overestimate or underestimate of the volume? Repeat with `rightsum`.

4. Compute the volume of the solid in Exercise 3 by integration and compare your answers.

5. *The volume of a JELL-O*^{TM} *mold.* Consider the region above the x-axis below the parabola that passes through the points $(8, 0)$, $(10, 4)$ and $(12, 0)$.

    (a) Find the volume of the solid obtained by revolving this region about the y-axis.

    (b) Find the volume of the solid obtained by revolving this region about the line $x = 2$.

    (c) Find the surface area obtained by revolving the parabola about the y-axis.

    (d) Find the surface area obtained by revolving the parabola about the line $x = 2$.

6. In this exercise, you are asked to calculate the approximate volume of a bowl of depth 3 feet with circular horizontal cross sections. The measurements of the radius of the cross sections versus height are given in the following table (in feet).

Height	0.00	0.50	1.00	1.50	2.00	2.50	3.00
Radius	0.00	0.55	1.05	1.40	1.70	1.85	2.00

    Find an approximate value of the volume of this bowl by adding the volumes of the disks of thickness 0.5 feet and radius given by the values in the table above.

7. Find a third degree polynomial whose graph contains every other data point in the table of the previous exercise. Use this polynomial to compute an approximate volume for the bowl. Compare your answer to that of the previous exercise.

    HINT: Think of the x-axis as representing depth, and enter $p := ax^3 + bx^2 + cx + d$ as an arrow-defined function. Solve four equations for the unknowns a, b, c and d so that the graph of p passes through the points $(0, 2)$, $(1, 1.7)$, $(2, 1.05)$, and $(3, 0)$. Graph this polynomial to check it contains these data points.

7.6. EXERCISES

8. Plot the Lissajous Figure

$$(x, y) = (\cos(pt), \sin(qt))$$

for $p = 3$ and $q = 2$ and find its arc length. Now plot the Lissajous Figure for $p = 6$ and $q = 4$ and find its arc length. Explain what you had to change to get the correct arc length.

9. A circular doughnut is formed by revolving a circle of radius b centered at $x = a$, $y = 0$ about the y-axis ($a > b$). Your job is to find the formula for the volume of this doughnut and evaluate it when $a = 4$ and $b = 3$.

   (a) In order to have the integral evaluate properly, you will need to communicate to Maple the assumptions that $a > b$ and $b > 0$. Use `assume(a>b,b>0);`

   (b) Find the equation of the circle described above for general a and b.

   (c) Review the cylindrical shell technique for finding volumes. Set up the integral for the volume of this doughnut and use Maple's `Int` and `value` commands to evaluate it. Check it using Pappas' Theorem which says this volume is equal to the product of the area of a circle or radius b and the circumference of a circle of radius a.

   (d) Evaluate the volume in the special case where $a = 4$ and $b = 3$.

   (e) To see a three-dimensional picture of this doughnut, issue the following Maple commands.
   ```
   >  x1:=(4+3*cos(t))*cos(s);
   >  y1:=3*sin(t);
   >  z1:=(4+3*cos(t))*sin(s);
   ```
   See if you can figure out why these three equations parameterize the doughnut as the parameters s and t vary from 0 to 2π. Then plot using
   ```
   >  plot3d([x1,y1,z1], s=0..2*Pi, t=0..2*Pi,
   >  scaling=constrained);
   ```

10. In this problem, your job is to find the formula for the surface area of the doughnut in the previous problem and evaluate it when $a = 4$ and $b = 3$.

    (a) Enter the assumptions that $a > b$ and $b > 0$.

    (b) Enter the following parametrization of the circle: $x = 4 + 3\cos t$, $y = 3\sin t$. Plot it to check it is a circle.

    (c) Enter the general parametrization: $x = a + b\cos t$, $y = b\sin t$ and compute the surface area of revolution for the doughnut for general a and b. Check it using Pappas' Theorem which says this area is equal to the product of the circumference of a circle or radius b and the circumference of a circle of radius a.

    (d) Evaluate the surface area in the special case where $a = 4$ and $b = 3$.

11. Generate the 5-th and 10-th order Fourier cosine expansions for $\sin(x)$ on the interval $[0, 7]$. Plot each of these against the function $\sin(x)$.

12. Repeat the previous exercise with $f(x) = 6 - x^2$ on the interval $[0, 4]$.

13. In Section 3.2, we plotted polar curves using the `polarplot` command from the `plots` package. We now want to compute the area of the region that lies between a polar curve $r = r(\theta)$ and the origin and between the angles $\theta = a$ and $\theta = b$. This area is given by

$$\frac{1}{2} \int_a^b r(\theta)^2 \, d\theta$$

Compute the area that is to the left of the parabola $r_1(\theta) = \dfrac{2}{1 - \cos(\theta)}$ and inside the circle $r_2(\theta) = 8$.

HINT: First plot the two curves to estimate the angles where they cross. Next use `fsolve` to find the angles more precisely. Then perform the appropriate integrals using the `Int` and `evalf` commands.

14. The arc length of a polar curve $r = r(\theta)$ for $a \leq \theta \leq b$, is given by

$$L = \int_a^b \sqrt{r(\theta)^2 + r'(\theta)^2} \, d\theta$$

and the surface areas of revolution of this polar curve about the x- and y-axes are, respectively

$$A = \int_a^b 2\pi y(\theta) \sqrt{r(\theta)^2 + r'(\theta)^2} \, d\theta$$

$$A = \int_a^b 2\pi x(\theta) \sqrt{r(\theta)^2 + r'(\theta)^2} \, d\theta$$

Here, $x(\theta) = r(\theta) \cos(\theta)$ and $y(\theta) = r(\theta) \sin(\theta)$.

(a) Plot the polar curve $r = 3 + \cos(\theta)$, for $0 \leq \theta \leq \pi/2$.

(b) Compute the arc length of the curve $r = 3 + \cos(\theta)$, for $0 \leq \theta \leq \pi/2$.

(c) Compute the area of the surface obtained by revolving about the x-axis the curve $r = 3 + \cos(\theta)$, for $0 \leq \theta \leq \pi/2$.

(d) Compute the area of the surface obtained by revolving about the y-axis the curve $r = 3 + \cos(\theta)$, for $0 \leq \theta \leq \pi/2$.

15. Plot the rose $r = \cos(5\theta)$. Compute the area of the surface obtained by revolving the petal for $-\pi/10 \leq \theta \leq \pi/10$, first about the x-axis and then about the y-axis. Be careful in choosing your limits of integration.

Chapter 8

Differential Equations

This chapter focuses on the use of Maple commands to analyze differential equations. We use Maple's `dsolve` command to find explicit solutions to various first and second order differential equations. Maple is very good at finding explicit solutions for differential equations—when they can be found. However, solutions to many of the important differential equations cannot be found in closed form. In this case, the `numeric` option of `dsolve` can be used to find approximate numerical solutions to differential equations. In addition, the `DEplot` command can be used to plot direction fields, solution curves and phase portraits.

8.1 Explicit Solutions

To solve the first order, linear differential equation $y' + 5y = 2x$, first enter the equation as

> `eq1:=diff(y(x),x)+5*y(x)=2*x;`

$$eq1 := (\tfrac{d}{dx}\,y(x)) + 5\,y(x) = 2\,x$$

Notice that you must write `y(x)` and not `y` so that Maple knows y is a function of x. Then solve the equation using `dsolve`:

> `dsolve(eq1,y(x));`

$$y(x) = -\frac{2}{25} + \frac{2\,x}{5} + e^{(-5\,x)}\,\_C1$$

Notice that the symbol `_C1` starts with an underscore, and so multiplies the $e^{(-5x)}$. This `_C1` is Maple's notation for the arbitrary constant that occurs in the general solution of a first order differential equation. Also notice that the output is an equation (rather than an expression or a function or an assignment). So you need to use the `rhs` command to assign the right hand side to a label:

> `ysol:=rhs(%);`

$$ysol := -\frac{2}{25} + \frac{2\,x}{5} + e^{(-5\,x)}\,\_C1$$

An initial condition such as $y(-1) = 2$ can be imposed in the following manner.

> `dsolve({eq1,y(-1)=2},y(x));`

$$y(x) = -\frac{2}{25} + \frac{2x}{5} + \frac{62\, e^{(-5x)}}{25\, e^5}$$

Note that the equation and the initial condition are enclosed in curly braces. To see a plot of the solution, the following Maple commands can be used:

> `ysol:=rhs(%):`
> `plot(ysol, x=-2..3, -1..4);`

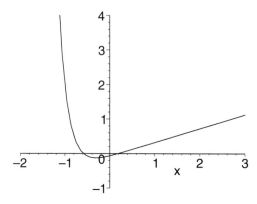

Notice that we have used the variables x and y here. However the independent variable is frequently taken as t to denote time.

The commands to solve a second (or higher) order differential equation are similar to the above. To find the general solution to the second order differential equation

$$\frac{d^2y}{dt^2} + 6\frac{dy}{dt} - 7y = 1$$

enter the commands

> `eq2:=diff(y(t),t$2)+6*diff(y(t),t)-7*y(t)=1;`

$$eq2 := (\tfrac{d^2}{dt^2}\, y(t)) + 6\,(\tfrac{d}{dt}\, y(t)) - 7\, y(t) = 1$$

> `dsolve(eq2,y(t));`

$$y(t) = e^t \, \_C2 + e^{(-7t)} \, \_C1 - \frac{1}{7}$$

Notice that there are two arbitrary constants in the solution, _C1 and _C2. To specify initial conditions such as $y(1) = 3$ and $y'(1) = 2$, enter

> `dsolve({eq2, y(1)=3, D(y)(1)=2},y(t));`

$$y(t) = \frac{3\,e^t}{e} + \frac{1}{7}\frac{e^{(-7t)}}{e^{(-7)}} - \frac{1}{7}$$

Notice that the `diff` command cannot be used to specify an initial condition in dsolve. Rather, you must use the D operator and function notation.

8.2 Direction Fields

When you can't solve a differential equation exactly, what do you do?

A direction field is a useful geometric device that aids in understanding the behavior of solutions to a first order differential equation, $\frac{dy}{dx} = F(x, y)$. At each point (x, y) on a solution $y = f(x)$, the slope of the solution curve is $m = f'(x) = F(x, y)$. The direction field is a way to display these slopes. At each point (x, y) draw a small line segment with slope $m = F(x, y)$. Then any solution curve will be everywhere tangent to these line segments.

Maple can plot direction fields using the DEplot command from the DEtools package. To plot the direction field for the differential equation $y' = x \sin(y)$, first load the DEtools package and define the equation:

```
> with(DEtools):
```

```
> deq:=diff(y(x),x)=x*sin(y(x));
```
$$deq := \tfrac{d}{dx} y(x) = x \sin(y(x))$$

Then use the DEplot command:

```
> DEplot(deq, y(x), x=-2..2, y=-1..1);
```

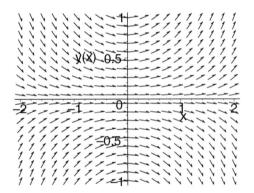

To graph one or more solutions to the differential equation together with the direction field, you need to add the initial conditions. Suppose you want three solution curves with initial conditions $f(0) = 0.5$, $f(1) = -0.5$ and $f(-1) = 0.25$. Then you execute

```
> inits:=[[0,.5], [1,-.5], [-1,.25]];
```
$$inits := [[0, 0.5], [1, -0.5], [-1, 0.25]]$$

```
> DEplot(deq, y(x), x=-2..2, y=-1..1, inits);
```

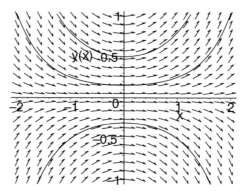

Notice how the solution curves are everywhere tangent to the direction field. In fact, this tangency is the basic idea behind Euler's method, which is a numerical algorithm for calculating approximate solutions to differential equations. Euler's method is discussed in Section 12.28.

Finally, to plot the curves without the direction field, add the option `arrows=none`:

```
> DEplot(deq, y(x), x=-2..2, y=-1..1, inits, arrows=none);
```

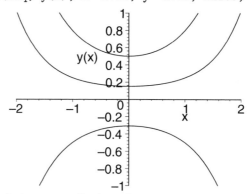

NOTE: If the solution curve displays sharp corners or other nonsmooth behavior, you should decrease the stepsize by inserting `stepsize=0.01` or whatever is appropriate as the last option in the `DEplot` command. The default step size is $1/20$ of the x-range, which is often too big.

8.3 Numerical Solutions

The material in this section is optional. However, since many nonlinear differential equations are virtually impossible to solve (in closed form), students should find the information in this section very useful.

Maple can find numerical approximations to solutions to differential equations. The algorithm that Maple uses is a Fehlberg fourth-fifth order Runge-Kutta method, which is more sophisticated than Euler's method. For example, suppose we want to solve $y' + \sin(y^2) = 1$, with $y(0) = 1$. If we proceed as in the previous section, we get the following.

8.3. NUMERICAL SOLUTIONS

```
>  deq:=diff(y(t),t)+sin(y(t)^2)=1;
```

$$deq := (\tfrac{d}{dt}\,y(t)) + \sin(y(t)^2) = 1$$

```
>  dsolve({deq,y(0)=1},y(t));
```

$$y(t) = \mathrm{RootOf}\left(t + \int_0^{-Z} \frac{1}{\sin(\_a^2) - 1}\,d\_a - \int_0^1 \frac{1}{\sin(\_a^2) - 1}\,d\_a\right)$$

Maple is telling us that it cannot evaluate some integral in closed form.

To obtain an approximate solution using dsolve, add the numeric option. (There must be a numeric initial condition.) Maple's output is then a procedure, which can be used in a manner similar to a Maple function. (For details on procedures, see Chapter 10.)

```
>  f:=dsolve({deq,y(0)=1},y(t),numeric);
```

$$f := \mathbf{proc}(x\_rkf45) \ \ldots \ \mathbf{end\ proc}$$

We have assigned the label f to this procedure so that we can use it to find approximate values for the solution at various values of t. We evaluate it just like a function. Its values at $t = 0$ and $t = 0.5$ are:

```
>  f(0); f(0.5);
```

$$[t = 0., \ y(t) = 1.]$$
$$[t = 0.5, \ y(t) = 1.06203222331694258]$$

To plot the solution we use Maple's odeplot command in the plots package:

```
>  with(plots):
>  odeplot(f,[t,y(t)],0..4, color=black);
```

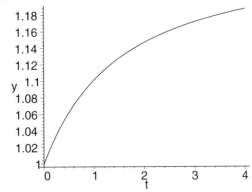

NOTE: The variables in the list $[t, y(t)]$ must be the same symbols for the independent and dependent variables that were used when the procedure f was defined. Alternatively, the plot can be obtained directly by using Maple's DEplot command in the DEtools package (possibly with less accuracy unless you decrease the stepsize) :

```
>  with(DEtools):
>  DEplot(deq,y(t),t=0..4,[[0,1]], arrows=none, stepsize=.01):
```

8.4 Systems of Differential Equations

One of the nicest features of Maple is its ability to analyze systems of differential equations. This section demonstrates the use of several Maple tools for analyzing systems of the form

$$x'(t) = f(x(t), y(t))$$
$$y'(t) = g(x(t), y(t))$$

For example, consider the Lotka-Volterra model given by

```
> de1:=diff(x(t),t)=x(t)*(1-y(t));
```

$$de1 := \tfrac{d}{dt} x(t) = x(t)\,(1 - y(t))$$

```
> de2:=diff(y(t),t)=0.2*y(t)*(x(t)-1);
```

$$de2 := \tfrac{d}{dt} y(t) = 0.2\, y(t)\,(x(t) - 1)$$

We start by loading the DEtools package.

```
> with(DEtools):
```

The DEplot command can be used to create a direction field plot as follows.

```
> DEplot([de1,de2], [x(t),y(t)], t=0..1, x=0..3, y=0..2,
> scaling=constrained);
```

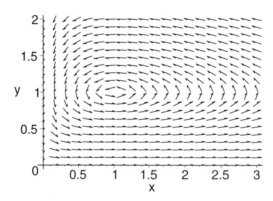

At each point, the small arrow shows the direction along which x and y will change as t increases.

We can add solution curves to the plot by adding initial conditions. Notice how the two sets of initial data are specified below.

```
> DEplot([de1,de2], [x(t),y(t)], t=0..20, [[x(0)=1.5,y(0)=1.5],
> [x(0)=2,y(0)=2]], stepsize=0.1, scaling=constrained);
```

8.4. SYSTEMS OF DIFFERENTIAL EQUATIONS

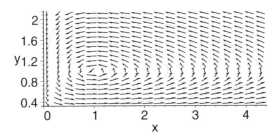

The solution curves are also called integral curves or phase trajectories, while the direction field plot together with several phase trajectories is called a phase portrait.

We conclude this section by demonstrating how Maple can be used to generate approximate numeric solutions to the system of differential equations. Suppose we want a solution to the Lotka-Volterra system given above corresponding to the initial data $x(0) = 1.5$ and $y(0) = 1.5$. We try to get an exact solution using dsolve

```
> sol:= dsolve({de1, de2, x(0)=1.5, y(0)=1.5}, {x(t),y(t)});
```

$$sol :=$$

but Maple can't do it, as indicated by the error message. So we use dsolve with the numeric option to obtain an approximate solution.

```
> sol:= dsolve({de1, de2, x(0)=1.5, y(0)=1.5}, {x(t),y(t)},
> numeric, output=listprocedure, abserr=0.001):
```

The result is a procedure. We can use this procedure to obtain approximations for our solution at any time value. For example, the approximate values at $t = 1$ and $t = 2$ are computed below.

```
> sol(1); sol(2);
```

$[t(1) = 1., \mathrm{x}(t)(1) = 0.873670307150112291, \mathrm{y}(t)(1) = 1.54975711402056548]$

$[t(2) = 2., \mathrm{x}(t)(2) = 0.524899401190241032, \mathrm{y}(t)(2) = 1.45335581162035842]$

We can also force Maple to extract only the portions that we are interested in. If we want separate functions for $x(t)$ and $y(t)$, then we can proceed as follows:

```
> xsol:=subs(sol,x(t)):   ysol:=subs(sol,y(t)):
> [xsol(1), ysol(1)];
```

$$[0.873670307150112291, 1.54975711402056548]$$

If we want to see a list of approximate values then we can create a loop.

```
> for i from 0 by 0.2 to 1 do
> [xsol(i), ysol(i)]
> end do;
```

$$[1.5000000000000, 1.5000000000000]$$

[1.35357000102293944, 1.52578802910667544]
[1.21617049705240476, 1.54322112219716723]
[1.08986906956533436, 1.55261211326470506]
[0.975711526594190914, 1.55452433804239254]
[0.873670307150112291, 1.54975711402056548]

It is also possible to plot the approximate solutions obtained from the `numeric` option of `dsolve` by using the `odeplot` command in the `plots` package. We first produce the phase trajectory:

```
> with(plots):
> odeplot(sol, [x(t),y(t)], t=0..20, numpoints=100,
> view=[0..3,0..2], scaling=constrained);
```

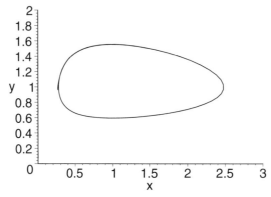

(This solution curve can also be produced using the `DEplot` command.) Notice, this plot shows the solution is periodic. To determine the period one can use trial and error on the range of t in the `odeplot` to see where the plot closes or one can separately plot $x(t)$ and $y(t)$:

```
> odeplot(sol, [[t,x(t)],[t,y(t)]], t=0..30, numpoints=100,
> view=[0..30,0..3]);
```

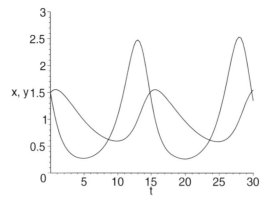

From this sketch, it appears as though the period of the solution is roughly 15.

8.5 Summary

- The dsolve command can be used to find exact solutions of single differential equations or systems of differential equations without or with initial conditions. The solution curves can be plotted using plot or the DEplot command, in the DEtools package.
- The dsolve command can also be used to find approximate solutions of single or systems of differential equations by adding the numeric option. The numeric solution curves can be plotted using the odeplot command, in the plots package.
- The DEplot command, in the DEtools package, can be used to plot direction fields, solution curves and phase portraits for single differential equations or systems of differential equations.

8.6 Exercises

In Exercises 1-3, use Maple to find the general solution of the differential equations. Then find and plot the particular solution satisfying the initial condition.

1. $\dfrac{dy}{dx} = \dfrac{e^x}{1+y}$ with $y(0) = 0$.

2. $y' + (\sin x)y = \sin(2x)$ with $y(\pi) = 1$.

3. $\dfrac{dy}{dx} = \dfrac{xy^2 + x}{x^2 + 1}$ with $y(2) = -1$.

4. Plot the direction field of the differential equation $y' = x^2 - y$ using DEplot. On a printout of this plot, hand sketch the solution of this differential equation that passes through the point $[0, 3]$. Then plot the solution and the direction field together. (Don't forget the *extra* set of square brackets surrounding initial conditions required by the DEplot command.)

5. Plot the direction fields for each of the differential equations in Exercises 1 - 3 together with the solution curve for the given initial condition.

6. Find the general solution of the differential equation: $y'' + y' + 5y = 0$. Then find the particular solution satisfying the initial conditions $y(0) = 1$ and $y'(0) = 2$. Finally plot the particular solution.

7. Solve the equation $y' + y = 2\sin(x)\cos(x)$, with initial condition $y(0) = 0$, both exactly and numerically. Use the display command to plot both solutions together using the range $-1 \leq x \leq 10$. How well does the numerical solution approximate the exact solution?

8. Use Maple to sketch the direction field for the system

$$\begin{aligned} x'(t) &= -0.1x(t) - y(t) \\ y'(t) &= x(t) - 0.1y(t) \end{aligned}$$

Sketch the solution trajectories associated with a variety of initial data pairs. What do you notice about the behavior of the solution curves?

9. The purpose of this problem is to describe the phase portraits for the system

$$x'(t) = ax(t) - by(t)$$
$$y'(t) = bx(t) + ay(t)$$

for a variety of choices of the parameters a and b.

(a) Start by setting $b = 1$. Plot the direction field for $a = -3, -2, -1, 0, 1, 2, 3$. What do these phase portraits have in common? How do they differ? Comment on whether a point moves in or out, clockwise or counterclockwise.

(b) Repeat this process with $b = -1$.

(c) Repeat this process with $b = 0$.

(d) Summarize your results.

10. Show that if c_1 and c_2 are constants then

$$x(t) = c_1 e^{at} \cos(bt) - c_2 e^{at} \sin(bt)$$
$$y(t) = c_1 e^{at} \sin(bt) + c_2 e^{at} \cos(bt)$$

solve the system in Exercise 9. Does this reinforce your conclusions from Exercise 9? Explain.

11. Consider the Lotka-Volterra system given by

$$x'(t) = x(t)(1 - y(t))$$
$$y'(t) = 0.2 y(t)(x(t) - 1)$$

The phase portrait for this system was sketched earlier in this chapter. From looking at the phase portrait, it appears as though the positive solutions of this system are periodic. Use the **numeric** option of **dsolve** to determine the period of the solutions to this system corresponding to each of the initial data in the following table:

$x(0)$	0.2	0.4	0.6	0.8	1.2	1.4	1.6
$y(0)$	0.2	0.4	0.6	0.8	1.2	1.4	1.6

Why was the initial condition $(1, 1)$ deliberately omitted from the table?
HINT: You might want to try a do loop:

```
>   for i from 0.2 by 0.2 to 1.6 do
>   inits:= x(0)=i, y(0)=i;
>   sol:=dsolve({deqs,inits}, {x(t),y(t)}, numeric);
>   plots[odeplot](sol, [x(t),y(t)], t=0..25);
>   end do;
```

Chapter 9

Sequences and Series

In this chapter, we use Maple to study sequences and series. In the first section, we define sequences using the `seq` command, plot them using the `plot` command and find their limits using the `Limit` and `value` commands. In the second section, we define series using the `Sum` command and compute their sums using the `value` command when the sum can be found exactly. When the sum cannot be found exactly, one must check for convergence and then estimate the sum. The third section shows how Maple's `Limit`, `Int` and `value` commands are useful in performing many convergence tests. When the series converges, it may be approximated by a partial sum of the series. Frequently, an integral can be used to bound the error in this approximation or to improve this approximation as shown in the fourth section. Finally in the fifth section we study Taylor series using the `taylor` and `TaylorApproximation` commands.

9.1 Sequences and Their Limits

To construct a sequence of numbers whose individual terms are given by a formula, first input the formula that describes the terms as a function of n.

> `a:=n->1/n^2;`

$$a := n \to \frac{1}{n^2}$$

To have Maple construct the first five terms of this sequence, use the `seq` command. Its first argument is an expression for the terms of the sequence. Its second argument tells Maple which terms to construct.

> `seq(a(n), n=1..5);`

$$1, \frac{1}{4}, \frac{1}{9}, \frac{1}{16}, \frac{1}{25}$$

To plot the sequence, you first need a list of points consisting of the index number and the corresponding term. Again, use the `seq` command

> `pointlist:=[seq([n,a(n)], n=1..5)];`

$$pointlist := [[1, 1], [2, \frac{1}{4}], [3, \frac{1}{9}], [4, \frac{1}{16}], [5, \frac{1}{25}]]$$

Notice the extra set of square brackets, making this a list of points as required by the `plot` command. (See Section 3.1 and the help on `?list`.) Now plot it:

> `plot(pointlist, 0..6, style=point, symbol=circle);`

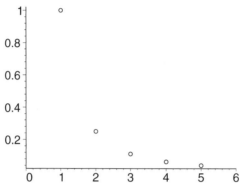

Often the main task with sequences is to determine the limiting value as n gets large. (This is denoted by $n \to \infty$ which is read as "n goes to infinity.") Maple is very good at calculating the limiting value but sometimes a plot gives a better idea of the sequence's behavior.

EXAMPLE: Examine the behavior of the sequence $b_n = \dfrac{2n-1}{3n+6}$ by plotting the sequence and by finding its limit as $n \to \infty$.

SOLUTION: Define and plot the sequence:

> `b:=n->(2*n-1)/(3*n+6);`

$$b := n \to \frac{2n-1}{3n+6}$$

> `pointlist:=[seq([n,b(n)], n=1..50)]:`
> `plot(pointlist, 0..50, 0..1, style=point, symbol=cross);`

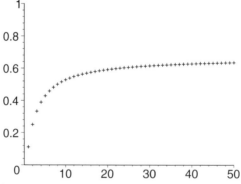

Notice that the `seq` command was terminated with a colon in order to suppress Maple's output. When first trying this, to check for typos, use a semicolon and have Maple display only the first few terms. Then click back on the line and edit it to get all desired terms and use a colon instead. In the `plot` command, if you leave off the option `style=point`, Maple will connect the points with straight lines

Maple's output is a plot of the first 50 terms of the sequence b_n. We see from the plot that as n goes to infinity, the sequence appears to have a limit which is slightly greater than 0.6. However, Maple can find the limit exactly, by using `Limit` and `value`.

> `Limit(b(n),n=infinity); value(%);`

$$\lim_{n \to \infty} \frac{2n-1}{3n+6}$$

$$\frac{2}{3}$$

which is slightly greater than 0.6. Recall that the `Limit` command (with a capital L) displays the limit in order to check for typos. The `value` command then evaluates it.

This sequence is so simple that Maple is not needed to compute the limit. (Divide numerator and denominator by n.) However this same procedure can be used to handle more complicated problems.

9.2 Series and Their Sums

A second operation commonly performed on a sequence a_n is to add its terms producing a series $\sum_{n=1}^{\infty} a_n$. (Note: the index of summation, n, does not need to begin with 1 and may end with a finite number or ∞.) To sum the terms, use Maple's `Sum` and `value` commands. If $a_n = \frac{1}{n^2}$, as in the previous section, then a finite sum might be

> `Sum(a(n),n=5..9); value(%);`

$$\sum_{n=5}^{9} \frac{1}{n^2}$$

$$\frac{737641}{6350400}$$

while an infinite sum is

> `Sum(a(n),n=1..infinity); value(%);`

$$\sum_{n=1}^{\infty} \frac{1}{n^2}$$

$$\frac{\pi^2}{6}$$

Notice that Maple is able to sum this series, while you are not expected to.

Maple may also be able to determine that a series diverges, for example:

```
> c:=n->1/n; Sum(c(n),n=1..infinity); value(%);
```

$$c := n \to \frac{1}{n}$$

$$\sum_{n=1}^{\infty} \frac{1}{n}$$

$$\infty$$

This infinite series is called the harmonic series.

Let's try one more, the geometric series:

```
> a:=n->r^n; Sum(a(n),n=0..infinity); value(%);
```

$$a := n \to r^n$$

$$\sum_{n=0}^{\infty} r^n \ .$$

$$-\frac{1}{r-1}$$

CAUTION: Unfortunately, Maple did not warn us that this series converges to the above value only if $|r| < 1$. For example, you might try to compute the value of the series $\sum_{n=0}^{\infty} 2^n$. So be careful!

Even though Maple sums the above series exactly, most infinite series cannot be summed exactly. So for many series, the goal is to discover whether or not the series converges, and to compute approximate values for its sum when it does converge.

9.3 Convergence of Series

There are four very useful tests for the convergence of a series of positive terms: ratio, root, limit comparison and integral tests. In this section, we will use Maple and these tests to determine whether or not a series of positive terms converges.

Ratio Test: Consider the series $\sum_{n=0}^{\infty} \frac{2^{3n}}{(2n+1)!}$. We input its terms as:

```
> a:=n->2^(3*n)/(2*n+1)!;
```

$$a := n \to \frac{2^{(3n)}}{(2n+1)!}$$

Again, the terms are entered as functions of n. This allows us to refer back to individual terms when we use the ratio or root test. Here the ratio test is appropriate. We compute:

```
> Limit(a(n+1)/a(n),n=infinity); value(%);
```

9.3. CONVERGENCE OF SERIES

$$\lim_{n\to\infty} \frac{2^{(3n+3)}(2n+1)!}{(2n+3)!\,2^{(3n)}}$$

$$0$$

Since the limit of the ratio of consecutive terms is 0, which is less than 1, the series converges.

Root Test: Similarly, for the root test, consider the series $\sum_{n=0}^{\infty} \frac{3^{2n}}{(2n+1)2^{3n}}$ which we input as:

```
> b:=n->3^(2*n)/(2*n+1)/2^(3*n);
```

$$b := n \to \frac{3^{(2n)}}{(2n+1)\,2^{(3n)}}$$

Using the root test, we compute:

```
> Limit(b(n)^(1/n),n=infinity); value(%);
```

$$\lim_{n\to\infty} \left(\frac{3^{(2n)}}{(2n+1)\,2^{(3n)}}\right)^{(\frac{1}{n})}$$

$$\frac{9}{8}$$

Since the limit is $\frac{9}{8}$ which is greater than 1, the series diverges.

When the limit in the ratio or root test is different from 1, we know whether the series converges or diverges. When this limit is 1, another test is needed. For example, for the series $\sum_{n=2}^{\infty} \frac{\ln(n)}{n^2}$ the limit of the ratios and the limit of the nth roots is 1. So we turn to another test.

Limit Comparison Test: Define the terms of the series $\sum_{n=2}^{\infty} a_n = \sum_{n=2}^{\infty} \frac{\ln(n)}{n^2}$

```
> a:=n->ln(n)/n^2;
```

$$a := n \to \frac{\ln(n)}{n^2}$$

Since $\ln(n)$ grows slower than any positive power of n, we can try a limit comparison test with the series $\sum_{n=2}^{\infty} b_n = \sum_{n=2}^{\infty} \frac{1}{n^{3/2}}$ which is convergent because it is a p-series with $p = \frac{3}{2} > 1$.

```
> b:=n->1/n^(3/2);
```

$$b := n \to \frac{1}{n^{(3/2)}}$$

```
> Limit(a(n)/b(n),n=infinity); value(%);
```

$$\lim_{n\to\infty} \frac{\ln(n)}{\sqrt{n}}$$

This says that the a_n's go to zero faster than the b_n's. So $\sum_{n=2}^{\infty} a_n = \sum_{n=2}^{\infty} \frac{\ln(n)}{n^2}$ also converges.

Integral Test: Again look at the series $\sum_{n=2}^{\infty} \frac{\ln(n)}{n^2}$. Since the terms are positive and decreasing, we can try the integral test.

> `a:=n->ln(n)/n^2;`

$$a := n \to \frac{\ln(n)}{n^2}$$

> `Int(a(n),n=2..infinity); value(%);`

$$\int_2^{\infty} \frac{\ln(n)}{n^2} \, dn$$

$$\frac{1}{2}\ln(2) + \frac{1}{2}$$

Since the integral is finite, the series $\sum_{n=2}^{\infty} \frac{\ln(n)}{n^2}$ converges.

9.4 Error Estimates

One of the more useful aspects of the integral test is its ability to estimate the error in using a partial sum to approximate a series. For example, we saw earlier that the sum of the series $\sum_{n=1}^{\infty} \frac{1}{n^2}$ is $\frac{\pi^2}{6}$. Suppose we didn't know this and added up the first 50 terms of the series. How close are we to the actual sum? Let's see:

> `a:=n->1/n^2;`

$$a := n \to \frac{1}{n^2}$$

> `Sum(a(n),n=1..50); S50:=evalf(%);`

$$\sum_{n=1}^{50} \frac{1}{n^2}$$

$$S50 := 1.625132734$$

> `err:=evalf(Pi^2/6-S50);`

$$err := 0.019801334$$

However, if we did not know the sum is $\frac{\pi^2}{6}$, could we find an upper bound on this error? Yes, using the ideas underlying the integral test! First note that the

9.4. ERROR ESTIMATES

error is

$$err = \sum_{n=1}^{\infty} \frac{1}{n^2} - \sum_{n=1}^{50} \frac{1}{n^2} = \sum_{n=51}^{\infty} \frac{1}{n^2}$$

This last sum is a lower Riemann sum for $\int_{50}^{\infty} \frac{1}{x^2} dx$ as illustrated below.

```
> rightbox(1/x^2, x=50..75, 25);
```

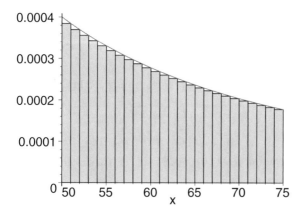

Consequently,

$$err = \sum_{n=51}^{\infty} \frac{1}{n^2} \leq \int_{50}^{\infty} \frac{dx}{x^2}$$

So we compute:

```
> Int(1/x^2,x=50..infinity); maxerr:=evalf(%);
```

$$\int_{50}^{\infty} \frac{1}{x^2} dx$$

$$maxerr := 0.02000000000$$

The error must be less than 0.02. We found it to be .019801334.

Integrals can also be used to improve the estimate. Note that the error

$$err = \sum_{n=51}^{\infty} \frac{1}{n^2}$$

is also an upper Riemann sum for $\int_{51}^{\infty} \frac{1}{x^2} dx$ as illustrated below.

```
> leftbox(1/x^2, x=51..76, 25);
```

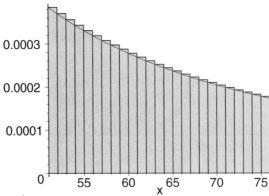

So we compute:

> `Int(1/x^2,x=51..infinity); minerr:=evalf(%);`

$$\int_{51}^{\infty} \frac{1}{x^2}\, dx$$

$$minerr := 0.01960784314$$

So the error must be at least .01960784314. Putting these facts together, the sum of the series $\sum_{n=1}^{\infty} \frac{1}{n^2}$ must be at least

> `Smin:=S50+minerr;`

$$Smin := 1.644740577$$

and at most

> `Smax:=S50+maxerr;`

$$Smax := 1.645132734$$

Thus the sum must be in the interval $[S_{min}, S_{max}]$. The midpoint of this interval is a better approximation to the sum:

> `Save:=(Smin+Smax)/2;`

$$Save := 1.644936656$$

The error in this new approximation is at most half the width of this interval.

> `Err:=(Smax-Smin)/2;`

$$Err := 0.0001960785$$

This is a significantly smaller error. Further, from the above we see that $\frac{\pi^2}{6} \approx 1.644937 \pm 0.000196$. Compare this to Maple's value

> `evalf(Pi^2/6);`

$$1.644934068$$

NOTE: This averaging method gives us a much better approximation of the infinite series than that obtained by just using the partial sum of the first 50 terms, as we did earlier. Moreover, the extra work involved is minimal.

9.5 Taylor Polynomials

An extremely useful idea in mathematics is the approximation of complicated functions with simpler ones. In Section 7.4 we studied Fourier cosine expansions which approximated functions by sums of cosines. In this section we study Taylor expansions, which approximate functions by polynomials.

Suppose we want the fifth degree Taylor polynomial of $\sin(x)$ about $x = 0$. The following Maple commands will construct this polynomial.

```
> taylor(sin(x),x=0,6); p:=convert(%,polynom);
```

$$x - \frac{1}{6}x^3 + \frac{1}{120}x^5 + O(x^6)$$

$$p := x - \frac{1}{6}x^3 + \frac{1}{120}x^5$$

A few words about the syntax are in order. Notice that the Maple command `taylor` has three parameters: the expression whose Taylor polynomial we want; the point that we expand about; and an integer. This last parameter—in this case 6—is one more than the degree of the desired Taylor polynomial. In Maple this number refers to the order of the error of the approximating polynomial. In the output, this error is denoted by $O(x^6)$. Notice too that the Maple command `convert(%,polynom)` drops the error term and enables us to assign the polynomial to a variable without having to retype it.

You can graph the function and its Taylor polynomial on the same plot:

```
> plot([sin(x),p],x=-2*Pi..2*Pi,-1.5..1.5);
```

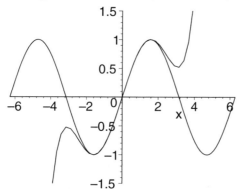

The higher the degree of the Taylor polynomial, the better the approximation.

Further, the `TaylorApproximation` command in the `Student[Calculus1]` package allows you to use a single command to compute a sequence of Taylor polynomials for a given expression:

```
> with(Student[Calculus1]):
> TaylorApproximation(sin(x), x=0, order=1..6);
```

$$x,\ x,\ x - \frac{1}{6}x^3,\ x - \frac{1}{6}x^3,\ x - \frac{1}{6}x^3 + \frac{1}{120}x^5,\ x - \frac{1}{6}x^3 + \frac{1}{120}x^5$$

or to plot them together with the original function:

```
> TaylorApproximation(sin(x), x=0, order=1..12, output=plot,
> -2*Pi..2*Pi, view=[-2*Pi..2*Pi,-1.5..1.5],
> functionoptions=[thickness=3], tayloroptions=[thickness=1]);
```

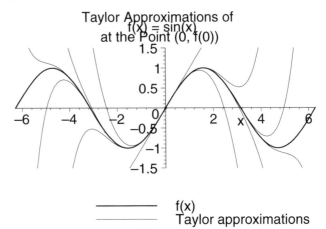

or animate them to see that the Taylor polynomials do approximate the original function.

```
> TaylorApproximation(sin(x), x=0, order=1..12,
> output=animation,-2*Pi..2*Pi, view=[-2*Pi..2*Pi,-1.5..1.5]):
```

Try this yourself, since you cannot see an animation in a book. To animate it, click in the plot and click on the PLAY button which is a triangle on the plot toolbar.

In addition, the Taylor Remainder Formula for the error term can often give a valuable estimate as to how well we have approximated our function. For example, suppose we want to approximate the sine function on the interval $[0, \pi]$ with its Taylor polynomial of degree 7, and we wish to know how good an approximation we have. First we decide to expand about the midpoint of the interval.

```
> taylor(sin(x),x=Pi/2,8):  p:=convert(%,polynom);
```

$$p := 1 - \frac{(x-\frac{\pi}{2})^2}{2} + \frac{(x-\frac{\pi}{2})^4}{24} - \frac{(x-\frac{\pi}{2})^6}{720}$$

The Taylor Remainder Formula says the remainder is

$$E_n = \sin(x) - p = \frac{f^{(n+1)}(c)}{(n+1)!}\left(x - \frac{\pi}{2}\right)^{(n+1)}$$

where c is some number between x and $\pi/2$. In this case, $n = 7$; so we need the eighth derivative of the sine function, evaluated at c. So we compute

```
> diff(sin(x),x$8); subs(x=c,%);
```

$$\sin(x)$$

$$\sin(c)$$

We know the absolute value of sine is never larger than 1. Moreover, the distance between x and $\pi/2$ is always less than or equal to $\pi/2$ (recall that x lies in the interval $[0, \pi]$). We can therefore estimate the error as follows.

$$|E_7| = |\sin(x) - p| = \left|\frac{f^{(8)}(c)}{8!}\left(x - \frac{\pi}{2}\right)^8\right| < \left|\frac{1}{8!}\left(\frac{\pi}{2}\right)^8\right|$$

> `evalf((Pi/2)^8/8!);`

$$0.0009192602758$$

Thus our seventh degree polynomial is uniformly within 0.00092 of the value of the sine function on the interval $[0, \pi]$. To visualize this, we plot the sine function together with the Taylor polynomial.

> `plot({sin(x),p},x=-Pi/2..3*Pi/2);`

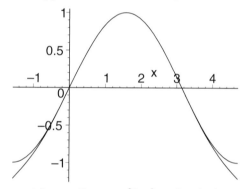

Notice the agreement is excellent on $[0, \pi]$ and only begins to vary near $-\pi/2$ and $3\pi/2$.

9.6 Summary

We studied sequences using the `seq`, `plot`, `Limit` and `value` commands. Then we studied series using the `Sum`, `value`, `Limit` and `Int` commands. Finally we studied Taylor series using the `taylor`, `convert(%,polynom)` and `TaylorApproximation` commands.

9.7 Exercises

In Exercises 1–7, plot the sequence. Try to determine whether or not the sequence has a limit as $n \to \infty$ and what the limiting value is. Then have Maple compute the limit if it exists.

1. $a_n = \dfrac{3n^3 - 6n^2 + 15}{2n^3 + 18n^2 - 6}$

2. $a_n = \dfrac{3n^3 - 6n^2 + 15}{2n^4 + 18n^2 - 6}$

3. $a_n = \dfrac{2n^4 + 18n^3 - 6}{3n^3 - 6n^2 + 15}$

4. $a_n = \dfrac{2 + (-1)^n n^2}{2n^2 + 3n + 4}$

5. $a_n = \dfrac{\ln(n^2)}{n^{2/3}}$

6. $a_n = \dfrac{n!}{20^n}$

7. $a_n = \left(1 + \dfrac{3}{n}\right)^{1/n}$

In Exercises 8–12 sum the given series using Maple. Then, simultaneously plot the first 10 terms of the sequence of terms a_n (In Maple: a(n).) and the first 10 terms of the sequence of partial sums $S_n = \sum_{k=1}^{n} a_k$. (In Maple: S:=n->sum(a(k),k=1..n).)

8. $\displaystyle\sum_{n=1}^{\infty} \dfrac{1}{4^n}$

9. $\displaystyle\sum_{n=1}^{\infty} \dfrac{(-1)^n}{n^2}$

10. $\displaystyle\sum_{n=1}^{\infty} \dfrac{1}{n^{k+1/2}}$, for $k = -1, 0, 1$

11. $\displaystyle\sum_{n=1}^{\infty} \dfrac{n^4}{3^n}$

12. $\displaystyle\sum_{n=1}^{\infty} \dfrac{1}{4 + n^2}$

In Exercises 13 and 14, decide whether the given series converges. In Exercises 15–18, give the values of x for which the series converges. For each of the series, try as many of the convergence tests discussed in this chapter as seem applicable.

13. $\displaystyle\sum_{n=0}^{\infty} \dfrac{n^3}{2n^5 + n^2 + 2}$

14. $\displaystyle\sum_{n=2}^{\infty} \dfrac{\ln^2(n)}{n \ln(n^{\ln(n)})}$

9.7. EXERCISES

15. $\sum_{n=1}^{\infty} \dfrac{(-1)^n x^n}{n\, 3^n}$

16. $\sum_{n=1}^{\infty} \dfrac{x^n}{e^n}$

17. $\sum_{n=2}^{\infty} \dfrac{(-1)^n x^{2n}}{n \ln(n)}$

18. $\sum_{n=1}^{\infty} \dfrac{(2x-3)^n}{n^p}$ for $p = \dfrac{1}{2}, 1, \dfrac{3}{2}$ and 2.

19. Find the sixth degree Taylor polynomial of $\cos(2x)$ about $x = \pi/3$. Then use the Taylor Remainder Formula to estimate the error in approximating $\cos(2x)$ with this Taylor polynomial on the interval $[0, \pi]$. Finally, plot both the function and the Taylor polynomial on the same coordinate axes.

20. Find the fourth through the eighth degree Taylor polynomials about $x = 1$ for $\dfrac{x^4 - 15x^2 + 2x - 5}{x^2 - 6}$. Plot all of them and the function on the same graph on the interval $[-1, 2]$. Then animate these plots. Finally use the Taylor Remainder Formula to estimate the error in approximating this function with the eighth degree Taylor polynomial on the interval $[-1, 2]$.

21. Find the 100th degree Taylor polynomial of $x^4 - 2x^2 + 15x - 6$ about $x = 0$ and also about $x = 6$, on the interval $[-2, 8]$. Explain why you should expect this result.

22. Estimate the value of the series $\sum_{n=1}^{\infty} \dfrac{1}{n^3}$ by using the sum of the first 50 terms of the series and by using the averaging method discussed in Section 9.4. Estimate the error in each approximation.

23. Calculate the seventh and eighth degree Taylor polynomials for the function $\cos(x)$ about $x = \pi/2$. The eighth degree polynomial should be more accurate than the seventh. Is it? How much extra work is involved in evaluating this eighth degree Taylor polynomial than in evaluating the seventh degree one? Next apply the Taylor Remainder Formula to estimate the error in each approximation on the interval $[0, \pi]$. Which error bound should you believe?

24. Compute the 5th degree Taylor polynomial centered at $a = 0$ for the function $f(x) = \sin(x)$. Name this polynomial $p(x)$ and evaluate $p(x^3)$. Now compute the 15th degree Taylor polynomial centered at $a = 0$ for the function $g(x) = \sin(x^3)$. What do you observe?

25. Compute the 6th degree Taylor polynomial centered at $a = 1$ for the function $f(x) = \tan(x)$. Name this polynomial $p(x)$ and evaluate $p(x^4)$. Now compute the 24th degree Taylor polynomial centered at $a = 0$ for the function $g(x) = \tan x^4$. What do you observe?

26. Use the previous 2 problems to make a conjecture concerning the relationship of the nth degree Taylor polynomial centered at $a = 0$ for a function $f(x)$ and the $(k \times n)^{th}$ degree Taylor polynomial centered at $a = 0$ for the function $f(x^k)$.

27. Compute the 5th degree Taylor polynomial centered at $a = 0$ for the function $f(x) = \cos(x)$. Name this polynomial $p(x)$ and evaluate $p(x^3)$. Now compute the 12th degree Taylor polynomial centered at $a = 0$ for the function $g(x) = \cos(x^3)$. What do you observe? In the problems above, we had the same sort of behavior, but the ratio of the Taylor orders would be 3:1 or 15:5. Why do we get the same result here with only a 12:5 ratio?

28. Use an appropriate degree Taylor polynomial centered at $a = 0$ for the function $f(x) = \sin(x^3)$ to approximate

$$\int_0^1 \sin(x^3)\,dx$$

to 15 decimal places of accuracy.

29. Repeat the previous problem for the integral

$$\int_{-3}^{3} e^{-x^2}\,dx$$

30. It is not always clear how to choose the value of the centering point for a Taylor polynomial. Consider the function $f(x) = e^{3x} + 7\sin(x)$ on the interval $[-1, 1]$. Let $p_{3,a}(x)$ denote the generic 3rd degree Taylor polynomial centered at $x = a$ for $f(x)$. Find the value of a so that

$$\int_{-1}^{1} [f(x) - p_{3,a}(x)]^2\,dx$$

is minimized.

Chapter 10

Programming with Maple

We have indicated in earlier chapters that Maple is a programming language and have already seen a simple do loop in applying Newton's Method in Chapter 4, Exercise 21. In fact, it is possible to use Maple to create highly structured programs that can perform a large number of useful tasks. Most importantly, any program written in Maple can call any of the high level commands available in Maple, or any program that you have previously written in Maple and included in the same session.

This chapter is intended as a short introduction to the Maple programming environment. This includes conditional and logical structures

```
if...then...elif/else...end if, and, or, not, true, false,
```
looping structures
```
for/while...do...end do, next, break
```
and procedural structures
```
proc...end proc, args, nargs, return, module.
```

10.1 Conditional and Logical structures

The first class of programming structures are conditional statements which tell Maple to perform certain commands provided a certain condition is satisfied.

EXAMPLE 1: Determine if the quadratic polynomial $3x^2 - 2x + 4$ has 2 distinct real roots.

SOLUTION: We know that $ax^2 + bx + c$ has 2 distinct real roots if the discriminant $b^2 - 4ac$ is positive. To compute this, we enter a polynomial:
```
>   p:=3*x^2-2*x+4;
```
$$p := 3x^2 - 2x + 4$$
read off the coefficients:
```
>   a,b,c:=coeff(p,x,2), coeff(p,x,1), coeff(p,x,0);
```

$$a, b, c := 3, -2, 4$$

and test the discriminant:
```
>    if b^2-4*a*c > 0
>    then "2 real roots"
>    else "not 2 real roots"
>    end if;
```
"not 2 real roots"

So the polynomial does not have 2 real roots.

Notice we entered the if...then...else...end if on multiple lines. You can do this by pressing ⟨SHIFT-ENTER⟩. If you just press ⟨ENTER⟩, the commands will work but you get a warning message that the input is incomplete. Putting the then and else clauses on separate lines is good style, but is not necessary.

In this example, we could have just computed the discriminant:
```
>    b^2-4*a*c;
```
$$-44$$

to see that it is negative. However, if this code is part of a larger computation and Maple needs to do something else based on the result of this test, then there might not be a human to decide if the discriminant is positive or negative.

"If" statements may be nested or additional conditions may be included by adding elif clauses. (This is an abbreviation for "else if.")

EXAMPLE 2: Determine if the quadratic polynomial $x^2 - 4x + 4$ has 2 real roots, 1 real root or 2 complex roots:

SOLUTION: We add an elif clause to our if statement:
```
>    q:=x^2-4*x+4;
```
$$q := x^2 - 4x + 4$$
```
>    a,b,c:=coeff(q,x,2), coeff(q,x,1), coeff(q,x,0);
```
$$a, b, c := 1, -4, 4$$
```
>    if b^2-4*a*c > 0
>    then "2 real roots"
>    elif b^2-4*a*c = 0
>    then "1 real root"
>    else "2 complex roots"
>    end if;
```
"1 real root"

Repeat this with the polynomials q:=x^2-3*x+4: and q:=x^2-5*x+4: to see the difference in output.

In general, the syntax for an "if" statement is:

> if *logical condition*
>   then *statements*
>   elif *logical condition*
>   then *statements*

10.1. CONDITIONAL AND LOGICAL STRUCTURES

```
      else statements
   end if;
```

Here, the `elif...then...` and `else...` clauses are optional and there may be an arbitrary number of `elif...then...` clauses. The *statements* following `then` or `else` can be any valid Maple commands, with each statement ending in either a colon or semicolon. The *logical conditions* following `if` or `elif` must evaluate to `true` or `false` and may be constructed using, for instance, the math and set relations:

Relation	Math Symbol	Maple Symbol
equal	$=$	=
not equal	$\neq$	<>
less than	$<$	<
less than or equal	$\leq$	<=
greater than	$>$	>
greater than or equal	$\geq$	>=
subset	$\subseteq$	subset
element of	$\in$	in

as well as the logical and set operators:

Operator	Math Symbol	Maple Symbol
and	$\wedge$	and
or	$\vee$	or
exclusive or		xor
not	$\neg$	not
implies	$\Rightarrow$	implies
union	$\cup$	union
intersection	$\cap$	intersection
set minus	$\setminus$	minus

The order of precedence for these operators may be found by typing `?precedence`, but to be safe *use parentheses* if you are unsure. You can also use an explicit `true` or `false`.

EXAMPLE 3: Given three numbers a, b and c, put them in ascending order and rename them x, y and z. Use two `if` statements: the first to put a and b in order and the second to put c in the correct position. Try it on $a = 7$, $b = -2$ and $c = 4$.

SOLUTION: Use intermediate variables m and n to denote a and b in ascending order:

```
>  a,b,c:=7,-2,4;
```
$$a, b, c := 7, -2, 4$$
```
>  if a<b
>  then m,n:=a,b;
>  else m,n:=b,a;
>  end if;
```
$$m, n := -2, 7$$

```
>    if c<m
>    then x,y,z:=c,m,n;
>    elif c<n
>    then x,y,z:=m,c,n;
>    else x,y,z:=m,n,c;
>    end if;
```
$$x, y, z := -2, 4, 7$$

Now try it with other choices of a, b and c.

10.2 Looping Structures

The second class of programming structures is looping structures (do loops) which tell Maple to perform certain commands repeatedly, with specified changes, until a certain condition is satisfied. You have already seen some examples of do loops in Chapter 4, Exercise 21 and Section 8.4.

In general, the syntax for a "do" loop is:

>   for *index* from *start* by *step* to *finish*
>   while *logical condition*
>   do
>   *statements*
>   end do;

Here, the *statements* between do and end do are executed repeatedly, with the variable *index* initially having the value *start* and incrementing each time by the amount *step* until either *index* becomes larger than the value *finish* or the *logical condition* in the while clause becomes false.

The for..., from..., by..., to... and while... clauses are all optional. If the from... or by... clauses are missing, then the values of *start* and *step* are both taken to have the default value 1. If the to and while clauses are both missing, then the loop will be repeated forever or until a break command is executed (as described below).

The *logical condition* in the while clause must evaluate to true or false and is constructed exactly like those in an if statement. The *statements* between do and end do can be any valid Maple commands, with each statement ending in either a colon or semicolon.

EXAMPLE 1: Find the first 4 derivatives of $f(x) = \sin(x)\cos(x)$ and see if you can find a pattern.

SOLUTION: Define the function and use a do loop to compute the derivatives:

```
>    f:=sin(x)*cos(x);
```
$$f := \sin(x)\cos(x)$$

```
>    for n to 4 do
>    diff(f,x$n)
>    end do;
```
$$\cos(x)^2 - \sin(x)^2$$

10.2. LOOPING STRUCTURES

$$-4\sin(x)\cos(x)$$
$$-4\cos(x)^2 + 4\sin(x)^2$$
$$16\sin(x)\cos(x)$$

NOTE: We entered multiple lines by pressing ⟨SHIFT-ENTER⟩. Further, we did not type a **from** or **by** clause; so by default, the index **n** starts at 1 and steps by 1 on each iteration.

Notice that $f^{(2)} = -4f$ and $f^{(3)} = -4f^{(1)}$. If n is even, so that $n = 2k$, then $f^{(2k)} = (-4)^k f$, and if n is odd, so that $n = 2k+1$ then $f^{(2k+1)} = (-4)^k f^{(1)}$.

Finally, there are two other commands which are used to control the flow in do loops: **next** and **break**. The **next** command causes Maple to skip the rest of that iteration and go on to the *next* iteration. The **break** command causes Maple to *break* out of the loop by skipping the rest of that iteration and all remaining iterations. Here is an example:

EXAMPLE 2: Find the first 10 derivatives of $f(x) = \sin(x)\cos(x)$. Evaluate each at $x = 0$. If a given derivative is positive at $x = 0$, also find its value at $x = \pi/2$. If the second evaluation is also zero, go on to the next derivative. If the second evaluation is negative, don't bother to compute any higher derivatives.

SOLUTION:

```
> for n to 10 do
> derf[n]:=diff(f,x$n);
> derf0[n]:=eval(derf[n],x=0);
> if derf0[n]<0
> then break
> elif derf0[n]=0
> then next
> end if;
> derfp[n]:=eval(derf[n],x=Pi/2)
> end do;
```

$$derf_1 := \cos(x)^2 - \sin(x)^2$$
$$derf0_1 := 1$$
$$derfp_1 := -1$$
$$derf_2 := -4\sin(x)\cos(x)$$
$$derf0_2 := 0$$
$$derf_3 := -4\cos(x)^2 + 4\sin(x)^2$$
$$derf0_3 := -4$$

Notice the first derivative is positive at $x = 0$, so Maple also computes the first derivative at $x = \pi/2$. The second derivative at $x = 0$ is zero, so Maple skips the second derivative at $x = \pi/2$. The third derivative at $x = 0$ is negative, so Maple skips the remaining derivatives.

Not all problems involve taking derivatives, and Maple can be applied to those problems as well. Here is an example.

EXAMPLE 3: **The Birthday Problem:** If n people are chosen at random, what is the probability that at least two of them have the same birthday? What is the smallest number of individuals necessary for the probability of a simultaneous birthday to be at least one half?

SOLUTION: We actually compute the probability that all the people have different birthdays and we gradually increase the number of people. If there are two people in a room, the likelihood that the second has a birthday different than the first is $\frac{364}{365}$. If neither of the previous individuals have the same birthday, then the likelihood that the third individual has still a different birthday is $\frac{364}{365} \times \frac{363}{365}$. It is clear how one would compute the probabilities for additional people, but the process of doing so is boring.

Such a problem is perfect for a Maple loop. We start with probability $p = 1.0$, (as a decimal to force all the results to be decimal) then repeatedly multiply by the respective factors while p is greater than one half. On the first iteration that the probability drops below the threshold, the loop ends. If the loop ends with a colon, the intermediate steps are not printed unless one uses a print statement.

```
> p:=1.0:
> for n from 2 while p>0.5 do
> p:=p*(366-n)/365:
> print(n,p);
> end do:
```

$$2, 0.9972602740$$
$$3, 0.9917958342$$
$$4, 0.9836440877$$
$$5, 0.9728644266$$
$$6, 0.9595375167$$
$$7, 0.9437642973$$
$$8, 0.9256647079$$
$$9, 0.9053761663$$
$$10, 0.8830518225$$
$$11, 0.8588586219$$
$$12, 0.8329752115$$
$$13, 0.8055897252$$
$$14, 0.7768974885$$
$$15, 0.7470986808$$
$$16, 0.7163959953$$
$$17, 0.6849923353$$
$$18, 0.6530885827$$
$$19, 0.6208814745$$
$$20, 0.5885616170$$

10.3 Procedures

$$21, 0.5563116655$$
$$22, 0.5243046929$$
$$23, 0.4927027663$$

This says that if there are 23 people in a room, there is only a 49% chance they all have different birthdays.

10.3 Procedures

Once you have written some code, you may want to reuse it with different input data. You don't want to have to retype the code every time with slight changes to the variable names or values. So you can write your own Maple commands with variable input (called parameters or arguments). These are called Maple programs or procedures. For example a procedure to add a first number to 3 times a second number is defined by

```
> mysum:=proc(a,b)
> a+3*b;
> end proc;
```

$$mysum := \mathbf{proc}(a, b)\, a + 3*b\, \mathbf{end\ proc}$$

It is called with a=2 and b=1 by executing

```
> mysum(2,1);
```

$$5$$

The arguments in the call can be expressions and are totally unrelated to the names of the parameters used in the definition of the procedure. For example:

```
> mysum(b,x+y);
```

$$b + 3x + 3y$$

Notice that parameter a in the definition has been replaced by the argument b in the call, and b has been replaced by x+y.

As an aside, you may be surprised to realize that you have already been writing simple Maple procedures in earlier chapters. These are the functions defined using arrow notation. You may be puzzled because they do not seem to have the form given above. However, the function defined by

```
> f:=(x,y)->x+3*y;
```

$$f := (x, y) \to x + 3y$$

is actually a shorthand for the procedure

```
> proc(x,y) x+3*y end proc;
```

$$\mathbf{proc}(x, y)\, x + 3*y\, \mathbf{end\ proc}$$

which is equivalent to the procedure mysum above. The arrow symbol tells Maple to create a procedure with parameters x and y and output $x + 3y$. They both produce the exact same output, e.g.

```
> f(2,1), f(b,x+y);
```

$$5,\ b+3x+3y$$

More generally, the syntax for a procedure is:

> *procname*:=proc(*parameters*)
>   local *variables*;
>   global *variables*;
>   *statements*
> end proc;

and it is called by executing:

> *procname*(*arguments*);

Here *procname* is the name you assign to the procedure. When you call the procedure, the values of the *arguments* in the parentheses become the values of the *parameters* in the definition; and then the *statements* are executed in order. The value returned by the procedure is *normally* the result of the last statement in the definition. (However, see the discussion of the return command below.) This returned value is the only thing printed out by the procedure, unless there are explicit print statements.

The local and global statements are optional. The *variables* in the local statement are only known within the procedure while the *variables* in the global statement are known both inside and outside the procedure. If you don't include these statements, Maple has complicated default rules for determining if the variables are local or global, and the result may not be what you intended. So it is best to include the statements, so that you are sure Maple is doing what you expect. The remaining *statements* in the procedure definition can be any valid Maple commands, with each statement ending in either a colon or semicolon.

Now let's create some procedures that are slightly more complex than the simple example given above. The first of these illustrates a use of the for/do looping structure.

EXAMPLE 1: In the previous section, we wrote a do loop to compute 4 derivatives of a function. Convert this into a procedure which will work for any function (written as an expression) and any number of derivatives. Apply the procedure to find the first 3 derivatives of the function $e^x \sin(x)$.

SOLUTION: First define the procedure

```
> derivs:=proc(f,k)
> local n;
> for n to k do
> print(diff(f,x$n))
> end do;
> end proc;
```

$derivs := \mathbf{proc}(f,\ k)\ \mathbf{local}\ n;\ \mathbf{for}\ n\ \mathbf{to}\ k\ \mathbf{do}\ \text{print}(\text{diff}(f,\ x\,\$\,n))\ \mathbf{end\ do\ end\ proc}$

NOTES: : We used a local variable n in the procedure so that executing the procedure does not change the value of any variable n used outside the procedure. Also, we explicitly typed print, since otherwise Maple would not print out intermediate results.

10.3. PROCEDURES

Now apply this program to the function $e^x \sin(x)$ and ask for 3 derivatives:

> `exp(x)*sin(x); derivs(%,3);`

$$e^x \sin(x)$$
$$e^x \sin(x) + e^x \cos(x)$$
$$2\, e^x \cos(x)$$
$$2\, e^x \cos(x) - 2\, e^x \sin(x)$$

NOTE: If the procedure ends with a semicolon, Maple will print the code so that one can check for errors. Once the procedure is working properly, there is seldom any reason not to suppress that output, and one can use a colon to clean up the worksheet as done below.

EXAMPLE 2: Write a procedure that will plot a function along with its tangent lines at several points. The inputs to the procedure should be the function, the two endpoints for the plot interval, and a list of the x-coordinates of the points of tangency. Use this program to plot $4\sin(x) + \cos(2x)$ along with its tangent lines at $x = -1.75,\ 2,\ 4$ on the interval $[-3, 6]$.

SOLUTION: The procedure (discussed below) is

```
> tan_plot:=proc(f,a,b,xs)
> local plist,x,i,xi;
> plist:=f(x);
> for i from 1 to nops(xs) do
> xi:=xs[i];
> plist:=plist,D(f)(xi)*(x-xi)+f(xi);
> end do;
> plot({plist},x=a..b);
> end proc:
```

and we apply it as follows:

> `f:=x-> 4*sin(x)+cos(2*x):`

> `tan_plot(f,-3,6,[-1.75,2,4]);`

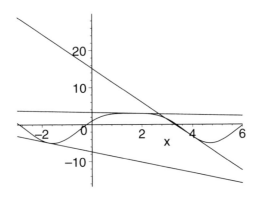

We need to discuss the details of the procedure. First, since the procedure refers to f(x) and D(f), we must define the function using arrow notation, not as an expression. Second, we used a new command **nops** (which stands for "number of operands"). This command returns the number of entries in a list (or, more generally, the number of operands in any operation). Consequently, the loop index i runs from 1 to the number of x's. The variable xi is defined as the ith x so we don't have to recompute it three times. (As a rule of thumb, if a quantity is recomputed more than twice, give it a name.)

The variable **plist** stores the functions to be plotted. Initially **plist** is just the original function. The loop then appends each of the tangent line functions, and the **plot** command encloses it in braces. Further, we defined all the variables to be local except for the arguments of the procedure.

Finally, note that, although there are several executable statements with semicolons, we only see the plot output from the last one: a Maple procedure only returns the value of the last executable statement.

One other feature of Maple procedures is that you do not need to know in advance of the procedure call the number of arguments that will be included. This is demonstrated in the next example.

EXAMPLE 3: Rewrite the procedure from example 2 so that the inputs are the function, the x-range and y-range for the plot, and the x-coordinates of the points of tangency. Use it to plot $x^3 - 3x^2$ along with its tangent lines at $x = 0, 1, 2, 3$ on the interval $[-1, 4]$.

NOTE: The x-coordinates are *not* put in a list; so you don't know in advance how many arguments there will be. Maple handles this with two new commands: **args** and **nargs**. Within the procedure, **nargs** is the number of arguments and **args** is the list of arguments.

SOLUTION: The procedure (discussed below) is

```
> tan_plot:=proc(f,xrange,yrange)
> local plist,x,i,xi;
> plist:=f(x);
> for i from 4 to nargs do
> xi:=args[i];
> plist:=plist,D(f)(xi)*(x-xi)+f(xi);
> end do;
> plot({plist},x=xrange,y=yrange);
> end proc:
```

and we apply it as follows:

```
> g:=x->x^3-3*x^2;
```

$$g := x \to x^3 - 3x^2$$

```
> tan_plot(g,-1..4,-10..10,0,1,2,3);
```

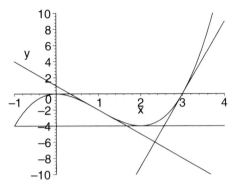

Again we need to discuss the details of the procedure. First, notice that when there are an indeterminate number of arguments, the first few are named in the procedure definition and the others remain unnamed. Second, the plot intervals are now given as ranges; so they must be entered in the form a..b instead of a,b which would count as two separate arguments. Third, instead of using nops(xs) we use nargs and instead of using xs[i] we use args[i]. Finally, the index i starts at 4, beginning after the three named arguments f, xrange and yrange.

A generalization of this example demonstrates the use of the return command to exit from the middle of a procedure and *return* a value which is not the result of the last statement in the procedure definition.

EXAMPLE 4: Rewrite the procedure from example 3 to check to see if any of the x-coordinates yield a vertical asymptote or vertical tangent. (A point $x = a$ is a vertical asymptote or tangent if $\lim\limits_{x=a} f'(x) = \pm\infty$.) If there is a vertical asymptote or tangent, print out an error message and return that value of x. Use the procedure to plot $\dfrac{x}{x-2}$, along with its tangent lines at $x = 1$, 2 and 3, on the interval $[0, 4]$. If there is a vertical asymptote or vertical tangent at one of the indicated points, drop that x-coordinate from the list and replot.

SOLUTION: The procedure (discussed below) is

```
> tan_plot:=proc(f,xrange,yrange)
> local plist,x,i,xi;
> plist:=f(x);
> for i from 4 to nargs do
> xi:=args[i];
> if limit(abs(D(f)(x)),x=xi)=infinity
> then print("Error:  The following point is a vertical
> asymptote or vertical tangent.");
> return x=xi;
> end if;
> plist:=plist,D(f)(xi)*(x-xi)+f(xi);
> end do;
> plot({plist}, x=xrange, y=yrange, discont=true);
> end proc:
```

and we apply it as follows:

> `h:=x->x/(x-2);`

$$h := x \to \frac{x}{x-2}$$

> `tan_plot(h,0..4,-10..10,1,2,3);`

"Error: The following point is a vertical asymptote or vertical tangent."

$$x = 2$$

Notice how Maple exits without completing the `do` loop or the remaining statements in the procedure. It did not print the result of the last statement, i.e., the `plot`. Rather it returned the value in the `return` statement, $x = 2$. Deleting 2 from the list of points, we get:

> `tan_plot(h,0..4,-10..10,1,3);`

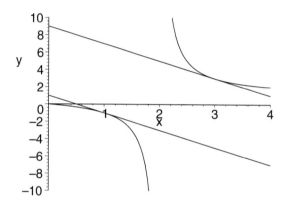

Here is another situation in which programming constructs are useful. Consider the Folium of Descartes, discussed in Section 4.4. Since y is not isolated in the equation, we must use implicit differentiation to compute the slope. We first write y as $y(x)$ to tell Maple that y is a function of x.

> `eq:=3*x*y=x^3+y^3:`
> `eq0:=eval(eq,y=y(x));`

$$eq0 := 3\,x\,y(x) = x^3 + y(x)^3$$

Then we take the derivative of both sides, and solve for y'.

> `diff(eq0,x);`
> `eq1:=diff(y(x),x)=solve(%,diff(y(x),x));`

$$3\,y(x) + 3\,x\left(\tfrac{d}{dx}\,y(x)\right) = 3\,x^2 + 3\,y(x)^2\left(\tfrac{d}{dx}\,y(x)\right)$$

$$eq1 := \tfrac{d}{dx}\,y(x) = \frac{y(x) - x^2}{-x + y(x)^2}$$

10.3. PROCEDURES

Now, suppose that we also want the second derivative or a higher derivative. To get y'' we differentiate the first derivative equation:

> `eq2:=diff(eq1,x);`

$$eq2 := \frac{d^2}{dx^2} y(x) = \frac{(\frac{d}{dx} y(x)) - 2x}{-x + y(x)^2} - \frac{(y(x) - x^2)(-1 + 2y(x)(\frac{d}{dx} y(x)))}{(-x + y(x)^2)^2}$$

However, we notice that the second derivative contains a y'. So we substitute the value of y' that we computed above and simplify.

> `lhs(eq2)=simplify(subs(eq1,rhs(eq2)));`

$$\frac{d^2}{dx^2} y(x) = -\frac{2 y(x) x (1 - 3 x y(x) + y(x)^3 + x^3)}{(-x + y(x)^2)^3}$$

To compute even higher derivatives, it is clear what needs to be done: at each stage, we differentiate the previous derivative and then replace y' by its value in terms of x and y. However, this manual process would quickly become onerous. Such a situation is exactly where one should consider a loop structure in a Maple program.

EXAMPLE 5: Write a Maple procedure which takes an equation in x and y, an x-range and y-range for a plot and an integer n, plots the equation and returns the nth derivative of y with respect to x. Use it to find y''' for the Folium of Descartes.

SOLUTION: Before writing the procedure, it is a good idea to first develop and test the central loop. Here it is for the third derivative. (Notice the use of embedded comments to help document the code. Anything following a sharp, #, is a Maple comment and is ignored when the line is executed.)

> `# Initialize eqi as the first implicit derivative.`
> `eqi:=eq1:`
> `# Compute the 3rd implicit derivative`
> `for i from 2 to 3 do`
> `diff(eqi,x):`
> `eqi:=lhs(%)=simplify(subs(eq1,rhs(%))):`
> `end do:`
> `print(eqi);`

$$\frac{d^3}{dx^3} y(x) =$$
$$\frac{2(1 - 3xy(x) + y(x)^3 + x^3)(-3y(x)x^2 - 4xy(x)^3 + x^4 + 5x^3 y(x)^2 + y(x)^5)}{(x - y(x)^2)^5}$$

We are now ready to embed this loop in a procedure. Notice the method used to obtain the first implicit derivative differs from the rest. So an `if` statement is used to determine which derivative is requested. When it is the first derivative, a `return` command is used to skip the rest of the procedure.

```
> implicitder:=proc(eq,xrange,yrange,n::posint)
> # Define local variables so any preexisting values
> # of i,eq0,eq1,eqi are unchanged
> local i,eq0,eq1,eqi;
> print(plots[implicitplot](eq, x=xrange, y=yrange,
> scaling=constrained, color=black));
> # replace y by y(x)
> eq0:=eval(eq,y=y(x));
> # Compute the first implicit derivative
> diff(eq0,x);
> eq1:=diff(y(x),x)=solve(%,diff(y(x),x));
> if n=1 then return simplify(eq1) end if;
> # Initialize eqi as the first implicit derivative
> eqi:=eq1;
> # Compute higher order implicit derivatives
> for i from 2 to n do
> diff(eqi,x):
> eqi:=lhs(%)=simplify(subs(eq1,rhs(%)));
> end do:
> eqi;
> end proc:
```

NOTE: One of the parameters is n::posint. This notation means that Maple should check the argument n to make sure it is a positive integer. If it is not, then an error message is generated. This powerful technique is called *type checking*. See ?type for a list of possible types.

Here is an example of the procedure at work:

```
> eq:=3*x*y=x^3+y^3;
> implicitder(eq,-2..2,-2..2,3);
```

$$eq := 3\,x\,y = x^3 + y^3$$

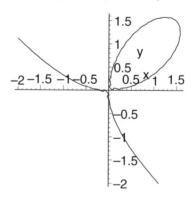

$$\frac{d^3}{dx^3}\,y(x) =$$

$$\frac{2\left(1 - 3\,x\,y(x) + y(x)^3 + x^3\right)\left(-3\,y(x)\,x^2 - 4\,x\,y(x)^3 + x^4 + 5\,x^3\,y(x)^2 + y(x)^5\right)}{(x - y(x)^2)^5}$$

NOTE: In the preceding procedure, we used the implicitplot command from the plots package. If we use the with(plots): command, then the whole package is read in and can affect things outside the procedure. So instead, we use the full name of the implicitplot command which is plots[implicitplot].

Everyone makes errors in writing computer programs. For help with debugging your Maple procedures, see Section 11.8.

Finally, we point out that once you have created a collection of related procedures, you can group them in a package by using the module command. See ?module for more details.

10.4 Additional Programming Constructs

We have given only a brief glimpse of the programming capability of Maple. We encourage you to explore all of Maple's programming capability by clicking on the HELP menu. In the Standard Interface, select MAPLE HELP and then from the TABLE OF CONTENTS select PROGRAMMING. In the Classic Interface, select INTRODUCTION and then from the first column of the Help Browser select PROGRAMMING. In either case, you can the explore the help pages on GENERAL INFORMATION, FLOW CONTROL, LOGIC, DATA TYPES, NAMES AND STRINGS, OPERATIONS, NOTATION, EXPRESSIONS, PROCEDURES AND FUNCTIONS, RANDOM OBJECTS, and INPUT AND OUTPUT.

10.5 Exercises

1. Enter the expressions $f = \dfrac{x^3}{x^2 + \sin(x)}$ and $g = \dfrac{x - \sin(x)}{x^2}$. Use an if statement to determine whether $f(3) > g(3)$, $f(3) = g(3)$ or $f(3) < g(3)$.

2. Write a do loop which will compute $\cos(1.0)$, $\cos(\cos(1.0))$, ..., $\cos(\cdots(\cos(1.0))\cdots)$ with 20 applications of cos. What do you think is the limit as the number of cos's becomes infinite?
   HINTS: Before the loop, execute x:=1.0. Then on each iteration execute x:=cos(%). Plot the curves $y = x$ and $y = f(x)$ on the interval $0..1$ and find the intersection using fsolve. With regard to the intersection point, what do you observe happening on alternate iterations of the loop? What method do you think Maple is using to find the intersection?
   NOTES: The final answer can also be obtained from (cos@@20)(1.0). Further, the process of finding the limit is known as finding a fixed point of the function.

3. In Section 10.1, Example 2 tests the sign of the discriminant of a quadratic polynomial to see if the polynomial has 2 real roots, 1 real root or 2 complex roots. A similar test can be done on cubic polynomials.
   For a general cubic polynomial $ax^3 + bx^2 + cx + d$, show that the change of coordinates from x to y where $x = y - \dfrac{b}{3a}$ will enable us to write the

polynomial in a simplified form with no y^2 term. Moreover, by dividing by a, we can make the polynomial have lead coefficient 1, so that it looks like $y^3 + py + q$.

The discriminant of $y^3 + py + q$ is $-4p^3 - 27q^2$. If the discriminant is positive, then there are three distinct real roots. If the discriminant is negative, there is one real root and two distinct complex roots. And if the discriminant is zero, then there are repeated roots.

Write a procedure whose argument is a cubic polynomial and which applies this test to its argument. Your procedure should print the original polynomial, the simplified polynomial, the value of the discriminant, the result of the test and the roots of the original cubic equation. Test it on the polynomials $x^3 - 6x^2 + 11x - 6$, $x^3 - 5x^2 + 8x - 4$ and $x^3 - 1$.

4. The first 10 Fibonacci numbers are $1, 1, 2, 3, 5, 8, 13, 21, 34, 55$. In general,

$$F_1 = 1, \qquad F_2 = 1, \qquad F_n = F_{n-2} + F_{n-1}$$

Write a procedure which returns the nth Fibonacci number.

5. Write a procedure which takes 2 numbers written in a Maple list and returns a list in either increasing order or decreasing order, depending on the second argument in the procedure.

6. Write a procedure which computes and simplifies the first n derivatives of an expression. Apply the procedure to find the first 4 derivatives of
$$\frac{e^x + e^{-x}}{e^x - e^{-x}}.$$

7. Write a procedure which computes the first n derivatives of a given function at a given number $x = a$. However, if one of these derivatives is zero at $x = a$, the procedure should quit after that derivative and print out an error message. Use either functions or expressions. If you use expressions, remember that the result of a `subs` command is not automatically simplified. So `simplify` the value of the derivative. Apply the procedure to compute the first 6 derivatives of

$$\frac{\sin(x)}{\cot(x) + \tan(x)}$$

at $x = \pi/3$ unless it is zero there. Then repeat the process with the same function but at $x = \pi/2$.

Chapter 11

Troubleshooting Tips

The Top Ten Headaches Reported by Beginning Maple Users:

1. **Missing or Incorrect Punctuation**
2. **Unmatched Punctuation and Commands**
3. **Confusion between Functions and Expressions**
4. **Unexpected Current Values of Variables**
5. **Using On-Line Help**
6. **Failure to Plot**
7. **Confusing Exact and Approximate Calculations**
8. **Debugging Procedures**
9. **Forgetting to Save a Maple Session and Losing It**
10. **Trying to Get Maple to Do Too Much**

This list is based on the actual experiences of many new Maple users, mostly freshman and sophomore mathematics students and their instructors. They brought with them a wide range of computer experience. However, Maple has its own idiosyncracies, and certain problems seem to arise, repeatedly, for just about everyone. It is hoped that by having these common errors pointed out, along with examples of how they occur, how to spot them, and tips for avoiding them, you will more quickly and painlessly achieve proficiency with this powerful software tool.

Before considering specific examples of common errors, we mention that many times such errors can be prevented by simply changing the font size at the beginning of a Maple session. This sounds trivial, but it's amazing how many students don't think about it, and suffer eyestrain and wasted time correcting errors caused by reading difficulties. In the same vein, many find it helpful to insert blank spaces in Maple statements or break lines at reasonable places and thereby improve readability. However, before you print your worksheet, be sure to check the PRINT PREVIEW and, if necessary, set your font back to a reasonable size, and resize any plots.

11.1 Missing or Incorrect Punctuation

Missing Semicolon or Colon at the End of a Line: The single most common error, and the easiest to fix, is forgetting the semicolon or colon at the end of a statement. In the Classic interface, you get the following error message:

```
>   f := x^2 + 1
```

```
Warning, premature end of input
```

Just backspace (twice), add the semicolon and press ⟨ENTER⟩! In the Standard interface, Maple gives a warning, but then inserts the missing punctuation and produces output anyway. Unfortunately, the warning messages make for a messy, hard to read document. To remove the warning, just click at the end of the line, add the semicolon and press ⟨ENTER⟩.

Missing Parentheses: Parentheses are used to group symbols in mathematical expressions, and failure to insert them properly can result in completely different expressions and unintended results. However, in order to correct the error, one needs to *see* the expression that Maple is evaluating.

Three techniques that allow one to see an expression before Maple executes it are using *inert* operators, using the last output % operator and assigning names and using them.

Suppose you want Maple to calculate $\sum_{k=1}^{10} \frac{1}{k(k+2)}$, and you enter

```
>   sum(1/k*(k+2), k=1..10);
```

$$\frac{19981}{1260}$$

The summand is not written correctly, so the sum is wrong. You might have caught the error by noticing that the output is too large. (The sum of ten numbers less than one cannot be greater than ten.) A better way to detect the error is to use the *inert* `Sum` and `value`:

```
>   Sum(1/k*(k+2), k=1..10); value(%);
```

$$\sum_{k=1}^{10} \frac{k+2}{k}$$

$$\frac{19981}{1260}$$

This makes it easy to spot the missing parenthesis, so that you can enter the corrected statement.

11.1. MISSING OR INCORRECT PUNCTUATION

```
>   Sum(1/(k*(k+2)), k=1..10); value(%);
```

$$\sum_{k=1}^{10} \frac{1}{k(k+2)}$$

$$\frac{175}{264}$$

Alternatively, you can use % to see how Maple interprets what you wrote, before it is evaluated in a sum:

```
>   1/(k*(k+1)); sum(%,k=1..10);
```

$$\frac{1}{k(k+1)}$$

$$\frac{10}{11}$$

or you can give the quantity a name and use it:

```
>   ak:=1/(k*(k+1)); sum(ak,k=1..10);
```

$$ak := \frac{1}{k(k+1)}$$

$$\frac{10}{11}$$

TIP: Blindly putting an expression that you have not seen into a Maple command, or putting one Maple command inside another Maple command is an invitation to disaster. Always assign the quantity a name, or use % or use a combination of an *inert* command, Sum, Limit, Diff or Int, to display the operation and then value(%) to compute it.

TIP: Try to get into the habit of pausing to examine the output from each line you enter, and asking yourself if it is reasonably close to what you were expecting. When writing a Maple procedure where you won't normally see the intermediate steps, either insert additional print statements or turn on tracing to help with debugging. (See Section 10.3.)

Incorrect Use of Brackets and Braces: Recall that in standard mathematical notation, square brackets and braces are often used like parentheses, as grouping symbols, especially when there are multiple levels of grouping. In Maple, however, square brackets and braces are *never* used as grouping symbols.

Square brackets define ordered lists, such as:

```
>   v:=[2,5,3,1,3,5];
```

$$v := [2, 5, 3, 1, 3, 5]$$

while a list of lists defines a matrix, such as:

```
>   A:=Matrix([[2,5,3,1,3,5],[4,1,2,6,2,1]]);
```

$$A := \begin{bmatrix} 2 & 5 & 3 & 1 & 3 & 5 \\ 4 & 1 & 2 & 6 & 2 & 1 \end{bmatrix}$$

On the other hand, braces define unordered lists or sets where the order of the entries is irrelevant and duplicate entries are ignored, such as

```
>  s:={2,5,3,1,3,5};
```
$$s := \{1, 2, 3, 5\}$$

If you don't use any delimiters, you get a sequence, such as
```
>  a:=2,5,3,1,3,5;
```
$$a := 2, 5, 3, 1, 3, 5$$

Sequences behave much like lists, except that a list counts as one argument in a function call, whereas a sequence of n expressions counts as n arguments in a function call.

Square brackets are also used to select entries, such as
```
>  v[4]; A[2,4]; s[2]; a[3];
```
$$1$$
$$6$$
$$2$$
$$3$$

In this context, square brackets can be used to produce subscripts, such as
```
>  u[2], b[n], g[1,3];
```
$$u_2, b_n, g_{1,3}$$

but Maple is actually trying to identify entries in undefined lists, sets, matrices or sequences.

However, we emphasize that square brackets and braces are *never* used as grouping symbols.

Missing * for Multiplication: Forgetting an asterisk, *, for multiplication can lead to errors which are not obvious. Here are some examples:
```
>  y := 2(x+1) + x^2;
```
$$y := 2 + x^2$$
```
>  y := (2+3)(3+5);
```
$$y := 5$$
```
>  y := (3+y)(x+2);
```
$$y := 8$$
```
>  y := x(y+2); whattype(%);
```
$$y := x(10)$$
$$function$$

In the first three examples, the second factor is apparently just ignored. In the fourth example, the second factor is treated as an argument of a function named x.

In fact, in all four examples, the first factor is treated as a function and the second factor is treated as its argument. This is clear in the fourth example, but what about the other examples? They also become clear when you learn that Maple regards a number as a constant function whose constant value is that number. So for example $2(x) = 2$ for all x.

11.1. MISSING OR INCORRECT PUNCTUATION

Using Decimal Points within Ranges: Suppose you want to plot the function x^2 on the interval $-.2 \leq x \leq .2$. *Notice the decimal points before the 2's.* If you type

```
>   plot(x^2,x=-.2...2);
```

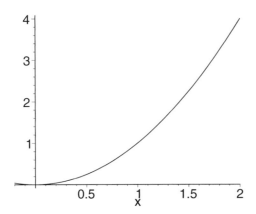

you get the plot on the interval $-.2 \leq x \leq 2$. The reason is that the symbol for a range is either .. or ... (or *any* number of consecutive periods) so that the last period is interpreted as part of the range symbol, rather than as a decimal point.

Consequently, the correct command to plot x^2 on the interval $-.2 \leq x \leq .2$ is (with zeros)

```
>   plot(x^2,x=-0.2..0.2);
```

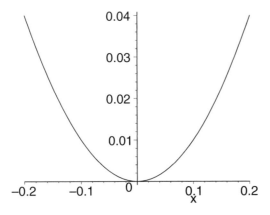

or (with a space)

```
>   plot(x^2, x=-.2..  .2);
```

which produces the same plot.

11.2 Unmatched Punctuation and Commands

Unmatched Parentheses, Brackets and Braces: Maple makes unmatched parentheses fairly easy to find, as the following example shows.

> `x*sqrt(2*x+1) / (x^2+5)*(x+1) );`

Error, ')' unexpected

Similarly, it finds unmatched square brackets and curly braces:

> `points:=[[1,2],[3,4,[5,6]];`

Error, ';' unexpected

> `sets:=[{1,2},{3,4,{5,6}];`

Error, ']' unexpected

By the way, it also finds a missing comma:

> `points:=[[1,2],[3,4][5,6]];`

Error, invalid subscript selector

although the error message is cryptic.

Forgetting `end if`, `end do` and `end proc`: The pairs `if...end if`, `do...end do` and `proc...end proc` act like parentheses. If you leave off the closing `end if`, `end do` or `end proc`, Maple complains: For example, if you type the following and press ⟨ENTER⟩,

> `a:=5;`

$$a := 5$$

> `if a<1 then a^2 + 1 else 3*a;`

Warning, premature end of input

you get a warning that something is incomplete. Just type `end if` and press ⟨ENTER⟩:

> `if a<1 then a^2 + 1 else 3*a;`
> `end if;`

$$15$$

The warning goes away and the answer appears. The same happens with incomplete `do` or `proc` statements.

TIP: If you need more than one line to type your input, pressing ⟨SHIFT-ENTER⟩ instead of ⟨ENTER⟩ will give a new line without executing the Maple code. This is useful for entering multiline commands, such as `if`, `do` and `proc`. Then, when you press ⟨ENTER⟩ without the ⟨SHIFT⟩, all of the lines will be executed.

Further, using ⟨SHIFT-ENTER⟩ to put your commands on multiple lines not only delays execution until the command is complete, but also make it easier to read, and hence, to debug your Maple code.

What happens when these commands are nested? Consider:

> `h := proc(x)`
> `if x<1 then x^2 + 1 else 3*x ;`
> `end proc;`

11.2. UNMATCHED PUNCTUATION AND COMMANDS

```
Error, reserved word 'proc' unexpected
```

This says Maple received an `end proc` when it expected (and required) an `end if`. The remedy is to insert the `end if` and press ⟨ENTER⟩ again.

```
> h := proc(x)
> if x<1 then x^2 + 1 else 3*x end if;
> end proc;
```

$$h := \mathbf{proc}(x) \, \mathbf{if} \, x < 1 \, \mathbf{then} \, x^2 + 1 \, \mathbf{else} \, 3 * x \, \mathbf{end} \, \mathbf{if} \, \mathbf{end} \, \mathbf{proc}$$

Notice that when a procedure is ended with a semicolon, one gets to see how Maple interprets the code. This is useful when you are debugging the code. However, once the procedure is complete and working, one should usually end the procedure with a colon to suppress the output, which is just a repetition of the input.

TIP: As soon as you type `if`, `do` or `proc` skip a few spaces or a few lines (by typing ⟨SHIFT-ENTER⟩) and type `end if`, `end do` or `end proc;` then back up and type what goes in between. This may help you to remember the `end if`, `end do` or `end proc`.

Unmatched Quotation Marks: Maple has three kinds of quotes, for three different purposes.

Name	Maple Symbol	Purpose
Double Quotes	"..."	for strings
Back Quotes	`...`	label for a variable name
Forward Quote	'...'	for delayed execution

Here are examples of the three kinds of quotes and the structures that they delimit:

Strings hold text you might want to display and are enclosed in double quotes (").

```
> "This is a string.";
```

"This is a string."

You can assign names to strings, manipulate them with Maple commands (for selecting substrings or concatenating strings) or print them with the `print` command. For more details, type `?string`. (Maple also has a `StringTools` package, if you want to get fancy.) For example:

```
> a:="notorbe";
```

$a := $ "notorbe"

```
> b:=cat(substring(a,3..4), " ", substring(a,6..7),
>   " ", substring(a,4..5), " ", substring(a,1..3),
>   " ", substring(a,3..4), " ", substring(a,6..7)):
> print(b);
```

"to be or not to be"

Names for variables normally must begin with a letter or an underscore (_) followed by other characters. To create a name that starts with a number or contains spaces or special characters, one must surround it by single left quotes, (`).

> `'3 * 4 is not':=7;`

$$3 * 4 \text{ is not} := 7$$

> `5+'3 * 4 is not';`

$$12$$

Forward single quotes (') delay the evaluation of a quantity for one execution cycle.

> `b,c:=2,3; a:='b+c'; a;`

$$b, c := 2, 3$$
$$a := b + c$$
$$5$$

Use them when you want to display the name of a variable instead of its value.

> `'b'=b;`

$$b = 2$$

but later still be able to evaluate it as its numeric value.

> `b;`

$$2$$

Alternatively, one can evaluate the variable as a name:

> `evaln(b)=b;`

$$b = 2$$

Here's another example where one might want to delay evaluation to improve readability:

> `p:=evalf(Pi,50):`
> `Int(1/x,x=0..p);`

$$\int_0^{3.1415926535897932384626433832795028841971693993751} \frac{1}{x} dx$$

If you want the integral to display the letter p but to compute with the value of π, then type:

> `Int(1/x,x=1..'p'); value(%);`

$$\int_1^p \frac{1}{x} dx$$

$$1.144729886$$

Finally, since forward quotes delay execution, 'p' is the name p instead of its value. Consequently, the statement

> `p:='p';`

$$p := p$$

assigns to p the name p (and not its value), thereby unassigning p.

If any of these three types of quotes are unmatched when you press ⟨ENTER⟩, Maple will complain.

```
>   "This is a string.;
```
Warning, incomplete string; use " to end the string
```
>   'This is a name.;
```
Warning, incomplete quoted name; use ` to end the name
```
>   '3*p;
```
Error, `;` unexpected

Likewise, you may have problems if you use the wrong kind of quotes.

This command is correct: a label is assigned as a name for a string.
```
>   'This is a name.':="This is a string.";
```
$$\textit{This is a name.} := \text{"This is a string."}$$

However, a string cannot be used as a label.
```
>   "This is a name.":=3;
```
Error, invalid left hand side of assignment
```
>   "b":=b;
```
Error, invalid left hand side of assignment

11.3 Confusion between Functions and Expressions

Definitions: Maple has two different mathematical constructs, expressions and (arrow-defined) functions. The natural one to work with is expressions. However, there are times when functions are more convenient.

An expression is simply a formula for computing something, for example:
```
>   f:=x^2+1;
```
$$f := x^2 + 1$$

On the other hand, a function is a rule which assigns a given output to a given input, for example:
```
>   g:=x -> x^2+1;
```
$$g := x \to x^2 + 1$$

As pointed out in Section 10.3, a function is actually a shorthand for a certain procedure. Thus, g above is totally equivalent to
```
>   g0:=proc(x) x^2+1 end proc:
```
Evaluation: Variables in expressions are given algebraic or numerical values with the subs or eval command:
```
>   subs(x=2,f); eval(f,x=2);
```
$$5$$
$$5$$

Functions are evaluated by putting the value of the argument in parentheses:

> `g(2);`

$$5$$

Errors occur when the syntax appropriate for a function is used with an expression or vice versa:

> `f(x);`

$$x(x)^2 + 1$$

> `f(1);`

$$x(1)^2 + 1$$

> `subs(x=1,g);`

$$g$$

Conversion: It is usually easy to convert expressions into functions, and vice versa:

> `f1 := unapply(f,x);`

$$f1 := x \to x^2 + 1$$

> `g1 := g(x);`

$$g1 := x^2 + 1$$

However, you should not use an arrow to convert an expression to a function; it will not evaluate properly. Recall that:

> `f;`

$$x^2 + 1$$

If we define the function

> `F:=x->f;`

$$F := x \to f$$

we might expect that `F(2)` would be 5. Instead we get:

> `F(2);`

$$x^2 + 1$$

Alternatively, you *can* use "copy and paste" to produce the function

> `F:=x->x^2+1;`

$$F := x \to x^2 + 1$$

However, there are times when you would not want to use "copy and paste", for example, when the formula is a very large result of a very long computation. You should also read Section 11.4 on the dangers in Using Copy and Paste.

Conversely, it may even be impossible to convert some more complcated functions into expressions. For example, if the function involves an `if` statement, such as

> `h:=x -> if x<1 then x^2+1 else 3*x fi:`

then it will not convert properly

> `h(x);`

11.3. CONFUSION BETWEEN FUNCTIONS AND EXPRESSIONS

```
Error, (in h) cannot determine if this expression is true or false: x
< 1
```
because Maple cannot determine if the variable x, not yet defined, satisfies the condition $x < 1$.

Plots: Maple treats functions and expressions differently in plots. With an expression in a plot command, the range *must* contain the plotting variable: `x=-10..10`. With an arrow-defined function or other procedure, the plotting variable x *must* be omitted: `-10..10`.

The following examples show correct ways of obtaining plots of the expression f and the function g, on both the default interval $-10 \leq x \leq 10$ and the interval $-2 \leq x \leq 2$. (We only show each plot once.)

```
> plot(f, x);
```

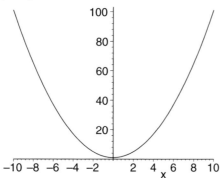

```
> plot(f, x=-2..2, 0..5);
```

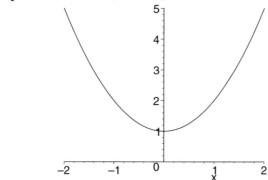

```
> plot(g);
```
(Plot not shown.)
```
> plot(g, -2..2, 0..5);
```
(Plot not shown.) For g, you can also use
```
> plot(g(x), x=-2..2, 0..5);
```
(Plot not shown.) since g(x) is an expression.

For h, which is a function and also has a discontinuity, you should use
```
> plot(h,-2..2, discont=true);
```

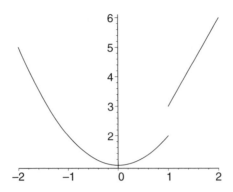

However, as pointed out above, the following shows that we can't plot h by converting it to the expression h(x), since h(x) isn't an expression.

```
> plot(h(x), x=-2..2, discont=true);
```

Error, (in h) cannot determine if this expression is true or false: x < 1

On the other hand, if you delay the evaluation of h(x) by surrounding it in single quotes, then it will execute because h(x) is not evaluated until after x is chosen:

```
> plot('h(x)', x=-2..2, discont=true);
```
(Plot not shown.)

Here are some more examples which are incorrect, since they mix the syntax for functions and expressions.

```
> plot(g(x), -2..2);
```

Warning, unable to evaluate the function to numeric values in the region; see the plotting command's help page to ensure the calling sequence is correct

Plotting error, empty plot

```
> plot(g, x=-2..2);
```

Error, (in plot) expected a range but received x = -2 .. 2

```
> plot(f(x), -2..2);
```

Warning, unable to evaluate the function to numeric values in the region; see the plotting command's help page to ensure the calling sequence is correct

Plotting error, empty plot

Finally, the following shouldn't work, but it does anyway.

```
> plot(f(x), x=-2..2);
```
(Plot not shown.)

11.4 Unexpected Current Values of Variables

Using the *!!!* Icon: When one first re-opens a Maple worksheet after it has been saved, the input and output shows in the worksheet, but none of it is in Maple's memory. To fix this you need to press ⟨ENTER⟩ on each line to re-execute it and put the result in memory. As a shortcut, you can click on the *!!!* icon in the toolbar, which automatically re-executes the entire worksheet in order from top to bottom.

A major problem with re-executing the worksheet can occur if you did not enter the commands in a linear order. In particular, suppose at some point in your computations you revise a portion of the worksheet near the top, based on values of a variable you computed near the bottom. Then you save, close and reopen the worksheet and re-execute the worksheet using the *!!!* icon. When the revised commands near the top are re-executed, they will not work properly because Maple does not yet know the values of the variables computed at the bottom. (A specific example is given below in the subsection on Re-executing Statements to Avoid Retyping.)

TIP: If you need to reuse some code from earlier in the worksheet with variables redefined later in the worksheet, copy and paste the code from earlier in the worksheet to down below where the variables have been redefined. Then re-execute them.

The *!!!* icon is also useful when Maple appears to misbehave in the middle of a session. In particular, the symbol % does not indicate the output of the command immediately preceding a given command, but instead, the output of the last command executed, regardless of where that command might lie. If one has moved about the worksheet to make corrections, it may well be the case that the last output is not the command above the command containing the %. Putting a `restart;` at the beginning of the worksheet and use of *!!!* will almost magically cure the resulting problem

TIP: Proper Maple procedure is always, if possible, to put the command containing the % on the same line (or in the same execution group) as the statement whose output it refers to. Students often neglect this important requirement.

Using COPY and PASTE: COPY and PASTE are available from the EDIT menu or from the keyboard as ⟨CTRL-C⟩ and ⟨CTRL-V⟩.

There is nothing wrong with using copy and paste to copy a region of a worksheet to another part of the worksheet, edit it and execute the modified code. The problem arises when you copy and paste the result of a computation instead of giving it a name and using the name, or using % to refer to the last output. Here is an example:

Suppose problem 4 in your book says to find the square of each solution of the equation, $x^2 - 7x + 4 = 0$. So you enter the equation, solve the equation and square each result:

```
>   eq:=x^2-7*x+4=0;
```
$$eq := x^2 - 7x + 4 = 0$$

```
>  fsolve(eq,x);
```
$$0.6277186767, 6.372281323$$
```
>  a:=(.6277186767)^2; b:=(6.372281323)^2;
```
$$a := 0.3940307371$$
$$b := 40.60596926$$

Notice we used copy and paste to copy each solution to the next line. At that point, you suddenly realize you did the wrong problem. You were supposed to do problem 5 which says to square each solution of the equation, $x^2 - 8x + 3 = 0$. So you change the equation and re-execute:

```
>  eq:=x^2-8*x+3=0;
```
$$eq := x^2 - 8x + 3 = 0$$
```
>  fsolve(eq,x);
```
$$0.3944487245, 7.605551275$$
```
>  a:=(.6277186767)^2; b:=(6.372281323)^2;
```
$$a := 0.3940307371$$
$$b := 40.60596926$$

Unfortunately, you now have the wrong answer! Copying and pasting results is dangerous! The better way to solve the problem is to give the answer a name and use it:

```
>  eq:=x^2-7*x+4=0;
```
$$eq := x^2 - 7x + 4 = 0$$
```
>  sol:=fsolve(eq,x);
```
$$sol := 0.6277186767, 6.372281323$$
```
>  a:=(sol[1])^2; b:=(sol[2])^2;
```
$$a := 0.3940307371$$
$$b := 40.60596926$$

That way, if you change the equation and re-execute, the other commands will automatically update.

```
>  eq:=x^2-8*x+3=0;
```
$$eq := x^2 - 8x + 3 = 0$$
```
>  sol:=fsolve(eq,x);
```
$$sol := 0.3944487245, 7.605551275$$
```
>  a:=(sol[1])^2; b:=(sol[2])^2;
```
$$a := 0.1555897963$$
$$b := 57.84441020$$

Re-executing Statements to Avoid Retyping: Re-executing statements can lead to errors because Maple variables may not be what you expect. For example, suppose you calculate the area of a circle of radius 2 by entering

```
>  r:=2;
```
$$r := 2$$

11.4. UNEXPECTED CURRENT VALUES OF VARIABLES

```
>   A:=Pi*r^2;
```
$$A := 4\pi$$
```
>   evalf(%);
```
$$12.56637062$$

Now suppose you want the area of a circle of radius 3. You enter

```
>   r:=3;
```
$$r := 3$$
```
>   evalf(A);
```
$$12.56637062$$

and get the wrong answer. The reason is that **A** hasn't been recalculated. If you now scroll up, click on the definition of **A** and press ⟨ENTER⟩, its output becomes

```
>   A:=Pi*r^2;
```
$$A := 9\pi$$

Now when you re-execute

```
>   evalf(A);
```
$$28.27433389$$

its output is correct.

However, there is still a problem. If you ever re-execute the worksheet, (Say you stopped in the middle one day and continued the next.) then the output would change to the wrong value. You *must* retype the definition of **A**.

TIP: To prevent re-execution errors, put several statements on one line. This ensures that they will all be re-executed when the line is re-entered.

```
>   r:=2; A:=Pi*%^2; evalf(%);
```
$$r := 2$$
$$A := 4\pi$$
$$12.56637062$$

TIP: To save time, use COPY and PASTE to copy the previous line to a new line, change the **2** to a **3** and press ⟨ENTER⟩.

```
>   r:=3; A:=Pi*r^2; evalf(%);
```
$$r := 3$$
$$A := 9\pi$$
$$28.27433389$$

Forgetting to Unassign Variables: Maple has a memory like an elephant.

As another example of Maple variables having unexpected values, supppose you have (unwisely) assigned a value to the letter **k**.

```
>   k:=3;
```
$$k := 3$$

Later, you want to calculate a sum and try to execute.

```
>   Sum(1/(k*(k+2)), k=1..10); value(%);
```

$$\sum_{3=1}^{10} \frac{1}{15}$$

```
Error, (in sum) summation variable previously assigned, second
argument evaluates to 3 = 1 .. 10
```

The problem is that the index of summation already has a value. Even worse you might just execute

> `sum(1/(k*(k+2)), k=1..10);`

```
Error, (in sum) summation variable previously assigned, second
argument evaluates to 3 = 1 .. 10
```

and not see the error in the sum. This error is easily corrected by unassigning k and making it a free variable again.

> `k := 'k';`

$$k := k$$

> `Sum(1/(k*(k+2)), k=1..10); value(%);`

$$\sum_{k=1}^{10} \frac{1}{k(k+2)}$$

$$\frac{175}{264}$$

Of course, if you have executed the assignment statement three problems before the one that you are working on now, it might be rather hard to recall that fact.

TIP: The best procedure is to perform a `restart:` command at the beginning of each new problem. This approach has the effect of clearing *all* previous assignment operations, and it limits how far back one might need to look for such assignments. (The one downside is that packages must be reentered, but that is a small price to pay.)

The following error is similar, but not as easy to diagnose. Suppose x is given a value

> `x:=1;`

$$x := 1$$

Then, at some later time, you decide to compute an integral:

> `int(x^2, x);`

```
Error, (in int) integration range or variable must be specified in the
second argument, got 1
```

The error message is very cryptic. Maple does not say the integration variable is previously assigned. The first step to understanding the problem is to switch to the *inert* `Int` command:

> `Int(x^2, x); value(%);`

$$\text{Int}(1, 1)$$

```
Error, (in int) integration range or variable must be specified in the
second argument, got 1
```

You can now see the problem is that x is replaced by a 1. (Moreover, you can check this by entering x; and observing the output.) Again, the remedy is to unassign x.

```
>   x:='x':  Int(x^2,x); value(%);
```

$$\int x^2\, dx$$

$$\frac{x^3}{3}$$

TIP: Avoid assigning anything to single letter variables, especially the variables x, y, z, t, i, j, k, m, or n so that they can be used for free variables in functions, plots, sums and integrals. Rather use variables such as x0, x1, i0, i1, etc. for values of the variables. (Alternatively, the subs command can frequently be used instead of assigning a value to the variable.)

11.5 Using On-Line Help

Maple has an excelent on-line help facility. Frustration with the on-line help is a common complaint of new users, even to the point that they stop trying to use it. This is unfortunate, because the on-line help is an extremely valuable resource in Maple, whether used as reference for looking up some forgotten syntax, or as a vehicle for exploring the myriad commands and mathematics Maple puts at your disposal.

Maple Help can be accessed either from the menubar, from the command line, by pressing ⟨F1⟩ or by pressing ⟨CTRL-F1⟩. The behavior is slightly different depending on whether you are using the Classic or Standard interface. The Standard interface is significantly better.

In the Classic interface, if you click on the HELP menu, you can access (i) the Maple Help Browser by clicking on INTRODUCTION, (ii) context help (i.e. help on the word currently containing the cursor), or (iii) the Topic Search or Full Text Search windows. You can also get to the context help directly by pressing ⟨F1⟩ or ⟨CTRL-F1⟩.

In the Standard interface, if you click on the HELP menu, you can access the Maple Help Browser by clicking on MAPLE HELP. This browser contains a box for a Topic Search or a Full Text Search. You access the context help by pressing ⟨CTRL-F1⟩. If you press ⟨F1⟩, you will get QUICK HELP on a very small number of common tasks.

In either interface, you can also access the help system from a prompt by typing 1, 2 or 3 question marks and the name of the command. (The number of question marks determines which part of the help page is emphasized.) In the Classic interface, this only works if you know the precise spelling of the command. In the Standard interface, this takes you to the Topic Search.

Once you are at the Topic Search box, if you type a word, Maple will offer a list of commands that deal with that word, including a definition from its dictionary. For example, if you type "implicit," then Maple will list commands like implicitdiff and implicitplot, plots, which you may click on for further information.

Once you are on a help page, the punch line of Maple Help, like the punch line of a joke, always occurs at the end, where there are always a collection of examples. (You can access the examples directly by typing 3 question marks and the name of the command.)

TIP: Always skip down to the examples. After you have studied the examples, you can copy and paste the commands from the example into your worksheet and then replace the sample data with your own data.

We will look at some typical situations.

Accessing Help from a Prompt: Suppose you want to plot two graphs on one coordinate system (you have seen it done before, so you know it's possible), but you don't remember the exact syntax. To see examples of the `plot` command, you enter

```
>   ???plot
```

It shouldn't take you very long to find the following example.

```
>   plot([sin(x), x-x^3/6], x=0..2, color=[red,blue],
>   style=[point,line]);
```

If you need more explanation of the command, enter

```
>   ?plot
```

for the full on-line help entry on `plot`, or else

```
>   ??plot
```

for an abbreviated entry just on the syntax of applying the command.

Use Help on a Need-to-Know Basis: Sometimes the explanations contain so much jargon and unfamiliar terminology that you're overwhelmed. It may be best to just ignore what you don't understand and not worry about it; you'll either learn it later, or else find you don't really need to know it. For example, suppose you want to find out about the `do` command, and you enter

```
>   ?do
```

You immediately encounter terms like "statement sequence" and "boolean expression", which may not be too meaningful to you, so you skip to the examples (which you could have gone to immediately with the `???do` statement), and find that these are quite sufficient for your purposes. In fact, if you're so inclined, you can use the examples to help you understand the explanation and its unfamiliar terms.

Similarly, Maple on-line help generally doesn't attempt to teach the mathematics that you may not understand, so you have to learn to separate the mathematics topics from the Maple topics. For example, the entry

```
>   ?int
```

will tell you about Maple's `int` and `Int` commands for evaluating *definite* and *indefinite integrals*, but it expects you to already know what these mathematical terms mean. In fact, the Standard interface now has a dictionary in which you can look up the definition of integral. You access it from the Topic Search under "integral, Definition".)

11.5. USING ON-LINE HELP

Sometimes something unexpected comes up, but can safely be ignored, as in the following example. You want to find a floating point decimal solution to the equation $e^x + x - 2 = 0$, and so you enter the following

```
>  solve(exp(x)+x-2=0,x);
```

$$-\text{LambertW}(e^2) + 2$$

"What in the world is LambertW?" you say to yourself. To find out, you decide to enter

```
>  ?LambertW
```

and see a screen full of information about something called Lambert's W function. Since you really only need a numerical approximation to the solution, you recall that you could have used `fsolve` instead of `solve`, and so you modify the above command and enter

```
>  fsolve(exp(x)+x-2=0,x);
```

$$0.4428544010$$

Alternatively, you could simply append `evalf(%)` after the `solve`. The point is that with so much information available in the on-line help, you should learn to selectively ignore what you don't need, while hunting down what you do need.

Help on Packages: Some Maple commands reside in packages that must be loaded into a Maple session before they can be used. To see a listing of the available packages, enter

```
>  ?index,package
```

Sometimes you may remember a command and its syntax, but have forgotten that it is part of a package. If you try to use the command before loading the package, not only won't it work, but the reason may not be readily apparent.

For example, suppose you want to plot a circle, and you remember using the `circle` command with arguments being the center and radius. So you enter:

```
>  circle([3,2],2);
```

$$\text{circle}([3, 2], 2)$$

Maple just repeats what you typed. This means the `circle` command is undefined, probably because it belongs to some package. So you look at its help.

```
>  ?circle
```

You see many choices of help pages, but select `circle, plottools` to see its help and also conclude that `circle` is in the `plottools` package. So you modify your input to read:

```
>  with(plottools):
>  circle([3,2],2);
```

$$\text{CURVES}([[5., 2.], [4.984229403, 2.250666467], \ldots$$

The result is just "CURVES" and a a lot of numbers. (NOTE: We deleted most of the output to save space.) So you look back at the examples on the help page and see the `circle` command requires the `display` command. To complete the problem, you modify your input to read:

```
> with(plots):
> display(circle([3,2],2), scaling=constrained);
```

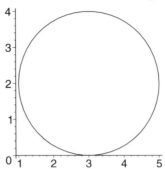

TIP: By the way, you only need to install a package once, not each time you use one of its commands. So if you use a package regularly, get in the habit of executing the `with` command at the top of the worksheet.

Spelling Problems: A similar problem arrises if you misspell a command. Maple will just repeat what you typed.

TIP: Remember that Maple is case sensitive. Some commands are all lower case, some are all upper case, some have the first letter capitalized, while others are in CamelCase (where words are run together, with capital letters to mark the start of each word). So it is easy to incorrectly capitalize a command. Spelling errors are even worse.

For example, suppose you remember that there is a command that will pop up a Tutor (a Maplet window) and allow you to play with approximate integration techniques. So you try

```
> ApproximateIntegralTutor(x^2,x=-2..2);
```
$$\text{ApproximateIntegralTutor}(x^2,\ x=-2..2)$$

Maple just repeats the command. So you try help:

```
> ?ApproximateIntegralTutor
```

And Maple can't find any help. So you try doing a Topic Search on ApproximateIntegralTutor or Approximate Integral Tutor or Integral Approximate to no avail. Finally you try doing a Full Text Search on Approximate Integral and find it. The correct name is ApproximateIntTutor and it is in the Student[Calculus1] package. So you enter

```
> with(Student[Calculus1]):
> ApproximateIntTutor(x^2,x=-2..2);
```

and get what you want.

TIP: When you are unsure of the spelling, try a Full Text Search which is available on the HELP menu in the Classic interface and at the top left of the Help Browser in the Standard interface with a radio button labeled TEXT.

TIP: Whenever Maple *parrots back* a command that you have typed, rather than executing it, you should check for missing packages and spelling errors.

11.6 Failure to Plot

Roots of Odd Degree: Students are often confused when they try to plot expressions containing roots of odd degree, like the cube root. For example, an excellent example of an expression that fails to have a derivative at the origin is the "Seagull" given by $y = x^{2/3}$. However, when one attempts to plot this expression a difficulty arises, and only half the plot is shown.

```
> x^(2/3); plot(%,x=-2..2);
```

$$x^{(2/3)}$$

The reason that the left hand side fails to appear is that Maple uses complex numbers. All real numbers have three cube roots, one real and two complex. When Maple computes the cube root of a positive number, it picks the real cube root. However, when it computes the cube root of a negative real number, it picks the first complex number counterclockwise from the positive real axis. Since Maple cannot plot the resulting complex number, no plot appears.

There are three easy solutions to this problem. One is to rewrite $x^{2/3}$ as $\left(x^2\right)^{1/3}$. A second is to replace the variable x by its absolute value `abs(x)`. And the third is to use the `surd` command. (See `?surd`.) Any of the three yields the desired full graph:

```
> plot((x^2)^(1/3),x=-2..2);
```

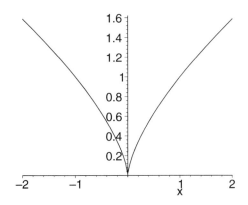

> `plot(abs(x)^(2/3),x=-2..2);`

> `plot(surd(x^2,3),x=-2..2);`

Unfortunately, for the very similar function $x^{1/3}$ you either need to use `surd` or plot the function in two parts:

> `plot(surd(x,3),x=-2..2);`

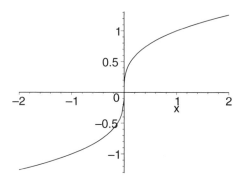

> `plot([x^(1/3), -(-x)^(1/3)],x=-2..2);`

Notice that the function $-(-x)^{1/3}$ is equivalent to $x^{1/3}$ but when x is negative, you are taking the cube root of the positive number $-x$.

Combining Plots with `display`: Yet another situation that students often encounter in which plots fail to appear is when one attempts to display two or more plots on the same pair of axes. One first creates individual plots, which are labeled a, b, etc, and ends the command with a colon, rather than a semicolon, to suppressing output. One can then enter `with(plots);` and the `display([a,b])` comand. However, it is not uncommon for one or more of the plots to fail to appear. Frequently the problem is that students often *create* the labeled plots, but never *check* that the plot actually *exists*. And in many cases it doesn't.

TIP: It is good Maple practice to show the plot at the same time that one creates and labels it. (This is frequently long before the `display` command.) This should be done by using the following syntax:

> `a:=plot(x^3,x):  a;`

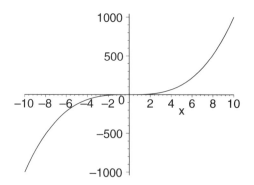

11.7 Confusing Exact and Approximate Calculations

Many commands work very differently depending on whether Maple is using exact arithmetic or floating point decimals. A decimal point in a single number is enough to cause Maple to use approximate arithmetic (decimals, rather than rational numbers). Consider the following examples.

Polynomials of degree four or less can be solved exactly using `solve`: (However, you may be less than happy with the result.).

> `solve(x^4+x-2=0, x);`

$$1, -\frac{(188 + 12\sqrt{249})^{(1/3)}}{6} + \frac{4}{3(188 + 12\sqrt{249})^{(1/3)}} - \frac{1}{3},$$

$$\frac{(188 + 12\sqrt{249})^{(1/3)}}{12} - \frac{2}{3(188 + 12\sqrt{249})^{(1/3)}} - \frac{1}{3}$$
$$+ \frac{1}{2} I \sqrt{3} \left(-\frac{(188 + 12\sqrt{249})^{(1/3)}}{6} - \frac{4}{3(188 + 12\sqrt{249})^{(1/3)}} \right),$$

$$\frac{(188 + 12\sqrt{249})^{(1/3)}}{12} - \frac{2}{3(188 + 12\sqrt{249})^{(1/3)}} - \frac{1}{3}$$
$$- \frac{1}{2} I \sqrt{3} \left(-\frac{(188 + 12\sqrt{249})^{(1/3)}}{6} - \frac{4}{3(188 + 12\sqrt{249})^{(1/3)}} \right)$$

or approximately by including a decimal point:

> `solve(x^4.+x-2=0, x);`

$1., 0.1766049821 + 1.202820819\, I, -1.353209964, 0.1766049821 - 1.202820819\, I$

Notice this is different from `fsolve` which returns only the real roots.

> `fsolve(x^4+x-2=0, x);`

$$-1.353209964, 1.$$

On the other hand, polynomials of degree five or higher can't in every case be solved exactly, and here's how Maple responds when asked for exact solutions to such equations. First an easy one:

> `solve(x^5-x=0, x);`

$$0, 1, -1, I, -I$$

and now a hard one:

> `solve(x^5+x-3=0, x);`

$\text{RootOf}(\_Z^5 + \_Z - 3, \text{index} = 1), \text{RootOf}(\_Z^5 + \_Z - 3, \text{index} = 2),$
$\text{RootOf}(\_Z^5 + \_Z - 3, \text{index} = 3), \text{RootOf}(\_Z^5 + \_Z - 3, \text{index} = 4),$
$\text{RootOf}(\_Z^5 + \_Z - 3, \text{index} = 5)$

Maple knows there are 5 roots (counting multiplicities) but it can't find them exactly using rational numbers. So it names them using the `RootOf` notation. However, Maple has a variety of ways for finding approximate roots, some of which we give below. In particular, the `evalf` command can be used to give a decimal approximation to a `RootOf` symbol.

> `evalf(%);`

$1.132997566,\ 0.4753807567 + 1.129701725\,I,\ -1.041879540 + 0.8228703381\,I,$
$-1.041879540 - 0.8228703381\,I,\ 0.4753807567 - 1.129701725\,I$

You can get directly to these solutions by including a decimal point (on the 3.)

> `solve(x^5+x-3.=0, x);`

$1.132997566,\ 0.4753807567 + 1.129701725\,I,\ -1.041879540 + 0.8228703381\,I,$
$-1.041879540 - 0.8228703381\,I,\ 0.4753807567 - 1.129701725\,I$

or you can get the real one(s) using `fsolve`:

> `fsolve(x^5+x-3=0, x);`

$$1.132997566$$

If all the roots of a polynomial — complex as well as real — are desired, the following will always work.

> `fsolve(x^5+x-3=0, x, complex);`

$-1.041879540 - 0.8228703381\,I,\ -1.041879540 + 0.8228703381\,I,$
$0.4753807567 - 1.129701725\,I,\ 0.4753807567 + 1.129701725\,I,\ 1.132997566$

Finally, we recall the example from Section 11.5, which also exhibits the difference between exact and approximate calculations.

> `solve(exp(x)+x-2=0, x); evalf(%);`

$$-\mathrm{LambertW}(e^2) + 2$$
$$0.442854401$$

11.8 Debugging Procedures

Maple procedures were discussed in Section 10.3. However, as with all programming, people make mistakes and need to debug their programs. We'll look at three types of errors: typing, syntax and math errors.

EXAMPLE 1: Write a program to find all critical points of a polynomial.
SOLUTION: You decide to write the polynomial as an expression and enter:

> `critpts:=proc(p)`
> `local eq,x,xsol;`
> `eq:=diff(p,x)=0;`
> `xsol:=fslove(eq,x);`
> `for x in [xsol] do`
> `print([x,subs(x,p)]);`
> `end do;`
> `end proc:`

11.8. DEBUGGING PROCEDURES

(We did not display the output because it is exactly a repeat of the input.) This procedure uses an alternate version of the do loop from that discussed in Section 10.2. Here, the loop is repeated with x taking on the value of each entry in the list [xsol]. (See ?do.) So it should print out the x- and y-coordinates of each critical point.

Now you try it on a polynomial you know has 4 zeros, hence 3 critical points.

```
> (x-2)*(x-3)*(x-4)*(x-5); q:=expand(%);
```
$$(x-2)(x-3)(x-4)(x-5)$$
$$q := x^4 - 14x^3 + 71x^2 - 154x + 120$$

```
> critpts(q);
```

Error, (in critpts) invalid input: subs received fslove(0 = 0,x), which is not valid for its 1st argument

Something went wrong. However, you look at the output and immediately notice that you misspelled fsolve. So you correct it. Now you have:

```
> critpts:=proc(p)
> local eq,x,xsol;
> eq:=diff(p,x)=0;
> xsol:=fsolve(eq,x);
> for x in [xsol] do
> print([x,subs(x,p)]);
> end do;
> end proc:
> critpts(q);
```
$$[x = x,\ x^4 - 14x^3 + 71x^2 - 154x + 120]$$

There is still something wrong with the fsolve command. This time there is confusion between the indeterminate x in the polynomial p and the variable name x in the procedure. By using a different variable name s instead of x, you now have

```
> critpts:=proc(p)
> local eq,s,xsol;
> eq:=diff(p,x)=0;
> xsol:=fsolve(eq,x);
> for s in [xsol] do
> print([s,subs(x=s,p)]);
> end do;
> end proc:
> critpts(q);
```
$$[2.381966011,\ -1.0000000]$$
$$[3.500000000,\ 0.5625000]$$
$$[4.618033989,\ -1.0000003]$$

It worked!

EXAMPLE 2: Write a procedure to sort three numbers in ascending order.
SOLUTION: The basic ideas were discussed at the end of Section 10.1. Here's the procedure and an execution:

```
> ascend:=proc(a,b,c)
> local m,n;
> if a>b
> then m,n:=a,b;
> else m,n:=b,a;
> end if;
> if c<m
> then c,m,n;
> elif c<n
> then m,c,n;
> else m,n,c;
> end if;
> end proc:
> ascend(4,7,-2);
```
$$-2,\ 7,\ 4$$

Something went wrong. This is probably a math error, and those are much harder to find.

Maple has a tool to help with debugging. At the beginning of the procedure, add the statement `option trace`.

```
> ascend:=proc(a,b,c)
> option trace;
> local m,n;
> if a>b
> then m,n:=a,b;
> else m,n:=b,a;
> end if;
> if c<m
> then c,m,n;
> elif c<n
> then m,c,n;
> else m,n,c;
> end if;
> end proc:
```

This will cause Maple to print out all the intermediate results as the procedure is executed:

```
> ascend(4,7,-2);
```

{--> enter ascend, args = 4, 7, -2

$$m,\ n := 7,\ 4$$
$$-2,\ 7,\ 4$$

<-- exit ascend (now at top level) = -2, 7, 4}

$$-2, 7, 4$$

You should immediately notice that m and n are backwards. So you look at the code and discover that a>b should be a<b. So you fix it and re-execute: (Don't forget to remove the trace once the procedure is working.)

```
> ascend:=proc(a,b,c)
> local m,n;
> if a<b
> then m,n:=a,b;
> else m,n:=b,a;
> end if;
> if c<m
> then c,m,n;
> elif c<n
> then m,c,n;
> else m,n,c;
> end if;
> end proc:
```

```
> ascend(4,7,-2);
```
$$-2, 4, 7$$

Another trick for debugging your programs is to add extra print statements along the way to print out intermediate results and let you know how far the program gets before the error occurs.

11.9 Forgetting to Save a Maple Session and Losing It

It sometimes happens that a Maple session will lock up and you have to kill it, or some other program you are running will die and you need to reboot your computer. If you are unable to save the session, you may lose a lot of work or be forced to start all over. For this reason, it's good practice to regularly save the session by clicking on FILE and SAVE. (There is also a SAVE icon that looks like a floppy disk on the toolbar. This button often works when clicking on the menu fails.)

Better yet, you can have Maple automatically save your file. See Appendix A for information on Maple's AutoSave feature.

11.10 Trying to Get Maple to Do Too Much

How to Interrupt a Computation: Whether by accident, miscalculation or poor judgement, a Maple computation sometimes takes longer than you expect. Suppose you want to compute $\sum_{n=0}^{200} n!$, but you mistakenly enter

> `Sum(n!, n=0..200000); value(%);`

Maple goes on forever. Try not to worry. Look at the tool bar and click on the red STOP sign (with a hand in the Standard interface). Maple should stop and give you the following warning.

> `Sum(n!, n=0..200000); value(%);`

$$\sum_{n=0}^{200000} n!$$

`Warning, computation interrupted`

Now you can correct your error and re-execute the command. Unfortunately, the stop sign doesn't always work. Let's look at another example.

Students often get carried away and ask Maple to do more than is reasonable. For example, you may have noticed that earlier in this chapter, we computed the first 50 digits of π by typing

> `evalf(Pi,50);`
    $3.1415926535897932384626433832795028841971693993751$

So you might ask if Maple can compute 100 or 500 or 5000 digits by just changing the 50. It can! Try it. But then you try 500,000 digits and Maple goes on forever!

> `evalf(Pi,500000);`

You try the STOP sign; and several things may happen:

- The computation stops. You get the warning message and you go merrily on your way.
- The computation does not stop. Further, the FILE muenu and SAVE icon are not available. Your only option is to close Maple by clicking on the X in the title bar or killing the task or process. At this point you may or may not be asked if you want to save your work. If this choice does not appear, then you have lost all the work you have not saved! That's why you need to *save your work often!*
- The computation does not stop. But worse, you get a curt pop-up window which says "mserver.exe has encountered a problem and needs to close." At this point you may or may not be asked if you want to save your work. If not, then you have lost all the work you have not saved! Again *save your work often!*

TIP: If you think a computation may take a long time, *save your work before executing the statement!*

TIP: You can set Maple to automatically save your files. See Appendix A.

11.10. TRYING TO GET MAPLE TO DO TOO MUCH

Maple Cannot and Does Not Need to Do Everything: We finally come to the section on what Maple cannot and should not do.

It frequently happens that, when students are given an assignment to use Maple on a problem, they proceed under the assumption that they must use Maple for every step in the problem. This can lead to the absurd situation of having to painstakingly cajole Maple into performing steps that are completely obvious, and should just be done by hand.

In addition, Maple cannot do the basic setup of the problem. Nor can it interpret the answer and judge if it is reasonable. Those are jobs for a human! Consider the following example:

EXAMPLE: A cylindrical aluminum can is to hold 500 cm^3 of soda. What are the dimensions (height and radius) of such a can which has the least surface area?

SOLUTION: Maple does not know that the volume of a cylinder is $V = \pi r^2 h$. Maple does not know that the surface area of a can is $A = 2\pi rh + 2\pi r^2$, where the first term is from the sides and the second term is from the two ends. Maple certainly cannot know that you need to include the two ends of the can in the surface area. In addition you have to tell Maple how to set up the problem and what steps to follow to solve it. Maple cannot read English nor interpret it as math. (Maybe sometime in the future, but not now.) That is your job.

So you proceed to enter the volume and set it equal to 500 and solve for h:

> Veq:=Pi*r^2*h=500;

$$Veq := \pi r^2 h = 500$$

> h1:=solve(Veq,h);

$$h1 := \frac{500}{\pi r^2}$$

These two steps are so obvious that you might just as well have done them in your head and entered:

> h1:=500/(Pi*r^2);

$$h1 := \frac{500}{\pi r^2}$$

Now you enter the area and substitute h1 for h:

> A:=2*Pi*r*h + 2*Pi*r^2;

$$A := 2\pi r h + 2\pi r^2$$

> A1:=subs(h=h1,A);

$$A1 := \frac{1000}{r} + 2\pi r^2$$

Next you set the derivative equal to zero and solve for r.

> DA:=diff(A1,r);

$$DA := -\frac{1000}{r^2} + 4\pi r$$

> r_sol:=fsolve(DA=0,r);

$$r\_sol := 4.301270069$$

Last you substitute r_sol back into h1.

> `subs(r=r_sol,h1); hsol:=evalf(%);`

$$\frac{27.02567690}{\pi}$$

$$hsol := 8.602540136$$

Finally, you interpret your solution: The optimal can has height 8.6 cm and radius 4.3 cm.

The point is Maple cannot set up the problem and does not know the steps to solve the problem, but it can do the algebra. In addition Maple does not need to do every step, if you can do it in your head. Finally, Maple cannot interpret the answer.

Chapter 12

Projects

This chapter contains a collection of projects on single variable calculus. Students may work individually or in groups of up to four students. Some projects may take one week. Others may require two or three weeks.

Projects for Calculus I

- A Power Relay Station
- The Search for the Meteorite
- Speed Limits
- Terminal Velocity
- The Ant and the Blade of Grass
- Distance Between Two Curves
- Tangent and Normal Lines
- Parameterizing Letters
- Parametric Curves
- Seeing a Blimp
- Static and Dynamic Tension

Projects for Calculus II

- Calculus I Review
- Calibrating a Dipstick
- The Area of a Unit p-Ball
- The Center of the State of Texas

- Center of Gravity of a Parabolic Plate
- The Oil Tank
- The Skimpy Donut
- Area Between a Curve and Its Tangent Line
- Curves Generated by Rolling Circles
- The Wankel Rotary Engine
- Shakespeare's Shylock
- The Bouncing Ball
- Pension Funds
- The Flight of a Baseball
- Parachuting
- Radioactive Waste at a Nuclear Power Plant
- Visualizing Euler's Method
- The Brightest Phase of Venus

12.1 A Power Relay Station

BACKGROUND: Review setting up word problems and graphing functions.

ASSIGNMENT: On a long straight highway, there are 11 communities located at mile markers $1.8, 2.2, 3, 3.35, 6.1, 7.35, 7.5, 8.3, 10.6, 11.2, 13.5$. The electric company which serves these 11 communities would like to construct a power relay station somewhere along this highway. The company would then run separate power lines from the relay station to each of the 11 communities. To minimize the cost of constructing these 11 power lines, the company would like to locate the relay station so as to minimize the sum of the distances to each station.

1. Think of the communities as located along the x-axis at the mile markers that are given. Suppose the relay station is located at position x. Write an expression for the sum of the distances from the power station to the communities using Maple's **abs** (absolute value) command. Plot it.

2. Where along the highway should the electric company build its relay station? Justify your answer. Is this location unique?

3. Suppose a 12th community, located at mile 15.7, is added to the highway. Now where should the electric company locate its relay station? Is this location unique?

12.2 The Search for the Meteorite

BACKGROUND: Review the equation of an hyperbola. You will use solve to find the intersection of two hyperbolas.

ASSIGNMENT: A meteorite crashes somewhere in the hills that lie north of point A. The impact is heard at point A and, 8 seconds later, it is heard at point B, 6 miles due west of A. Six seconds still later it is heard at point C, 4 miles due east of A. Locate the point of impact of the meteorite. The speed of sound is roughly .20 miles per second. Include an accurate plot with your explanation.

HINT 1: The equation of the hyperbola centered at the origin with focal points at $(+c, 0)$ and $(-c, 0)$ along the x-axis, is given by

$$\frac{x^2}{a^2} - \frac{y^2}{c^2 - a^2} = 1$$

Here, $2a$ is the difference $|d_1 - d_2|$, where d_1 is the distance between an arbitrary point (x, y) on the hyperbola and the focal point $(+c, 0)$, and d_2 is the distance between (x, y) and the focal point $(-c, 0)$.

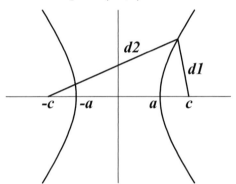

HINT 2: For this project, locate points A, B, and C on the x-axis. Place the origin at the midpoint between points A and B. From the information given, the meteorite must lie on an hyperbola centered at the origin with points A and B as focal points. Find the equation of this hyperbola. Likewise, the meteorite must also lie on an hyperbola with focal points at A and C. (Where is this hyperbola centered? What is its equation?) Thus, the meteorite must lie at the point of intersection of the two hyperbolas.

NOTE: An hyperbola has two branches. You must take care to choose the correct branch of each hyperbola.

12.3 Speed Limits

GOAL: The goal of this project is to learn about average and instantaneous velocities. In the process you will use Maple to define, evaluate and plot functions and compute limits.

BACKGROUND: Before you start, try each of the following examples.

- Definition and evaluation of functions.
  The following Maple command defines the function $x(t) = \dfrac{t^3 - 9t}{3t^3 - 9t^2 + t - 3}$, which might be interpreted as the position of an object at time t.
  ```
  >   x:=t -> (t^3-9*t)/(3*t^3-9*t^2+t-3);
  ```
  Then the position at $t = 1$ is
  ```
  >   x(1);
  ```
  You can also define functions of two or more variables. For example, the following gives the average velocity of the object between the times $t = t_1$ and $t = t_2$.
  ```
  >   AveVel:=(t1,t2) -> (x(t2)-x(t1))/(t2-t1);
  ```
  Then the average velocity between $t = 4$ and $t = 5$ is
  ```
  >   AveVel(4,5);
  ```

- Plots.
  The function $x(t)$ defined above, can be plotted on the interval $0 \le t \le 10$ by using the Maple command
  ```
  >   plot(x(t),t=0..10);
  ```

- Limits.
  Notice that the function $x(t)$ is undefined at $t = 3$ because the denominator is zero there. Maple agrees.
  ```
  >   x(3);
  ```
  Error, (in x) numeric exception: division by zero

  However, Maple can compute the limit as t goes to 3.
  ```
  >   Limit(x(t),t=3);   value(%);   evalf(%);
  ```
  You should be able to verify this limit, both in the plot and by algebraically manipulating the formula for $x(t)$.

ASSIGNMENT: A Texas Aggie suddenly decided to take his girlfriend to Houston for the day. So he hopped in his car and started to drive from College Station to Houston. When he got to Navasota, he suddenly realized that he had forgotten his girlfriend. Embarrassed, he slowly drove back to College Station and got his girlfriend. All this took 2 hours. Then they proceeded to Houston, taking another 2 hours. His distance (in miles) from College Station at time t (in hours) was $x(t) = \dfrac{200t^2(t-2)^2}{3+9t^2}$.

1. Enter his position function $x(t)$ into Maple using an arrow definition.

2. Plot his position $x(t)$ for times between $t = 0$ when he first left College Station and $t = 4$ when he finally arrived in Houston.

3. Approximately when did he arrive in Navasota? Approximately, what is the distance from College Station to Navasota?
   HINT: Click in the plot window at the first maximum and read off the t and x coordinates (at the left end of the toolbar).

4. What is the exact distance from College Station to Houston?
   HINT: Compute $x(4)$.

5. Find his average velocity (Use `AveVel` defined above.)
   (a) from College Station to Navasota, the first time out. (approximate).
   (b) from Navasota back to College Station. (approximate).
   (c) from College Station to Houston, the second time out. (exact).
   (d) from College Station to Houston, the whole trip. (exact).

   NOTE: To avoid round-off error in the next problem, execute
   > `Digits := 20;`

6. Find his average velocity between time $t = 3$ and time $t = 3 + h$ for the following values of h

1	.1	.01	.0001
-1	$-.1$	$-.01$	$-.0001$

7. Use the values calculated above to guess his instantaneous velocity at $t = 3$.

8. (Clear h by executing `h:='h';`) Compute his average velocity between time $t = 3$ and time $t = 3 + h$ for a variable value of h. Use this formula to compute his instantaneous velocity at $t = 3$ by computing the limit
$$\lim_{h \to 0} AveVel(3, 3 + h)$$

9. Recompute the limit above by algebraically manipulating $AveVel(3, 3+h)$ using the `expand` and/or `factor` commands until the h in the denominator cancels. Then substitute in $h = 0$ using the `subs` command.

10. *10% extra credit.* At what time does he pass through Navasota on his second time out? HINT: If X is the distance to Navasota, solve the equation $x(t) = X$ using Maple's `fsolve` command.

12.4 Terminal Velocity

GOAL: The goal of this lab is to learn about asymptotic (or terminal) velocity.
BACKGROUND: You will use Maple to define, evaluate and plot expressions and compute limits at infinity. Before you start, try each of the following examples.

- Definition and evaluation of expressions.
  The following Maple command defines the expression $x = \dfrac{t^3 - 9t}{3t^3 - 9t^2 + t - 3}$, which might be interpreted as the position of an object at time t.
  ```
  > x:=(t^3-9*t)/(3*t^3-9*t^2+t-3);
  ```
  Then the position at $t = 1$ is
  ```
  > eval(x, t=1);
  ```

- Plots.
  The expression x defined above, can be plotted on the interval $0 \le t \le 10$ by using the Maple command
  ```
  > plot(x,t=0..10);
  ```

- Limits.
  Notice that the expression x appears to approach a value of about $x \approx .5$ as t gets large. Maple can compute this limit precisely:
  ```
  > Limit(x,t=infinity);   value(%);
  ```
  Notice the limit is not .5! This is because the plot did not look far enough out in t. Correct the plot by changing the range for t. You should also be able to verify this limit by algebraically manipulating the formula for x.

ASSIGNMENT: A parachute jumper's velocity is $v = \dfrac{100t + 85t^2}{1 + 23t^2}$ in meters per second where t is in seconds.

1. Enter her velocity v into Maple as an expression.

2. Plot her velocity v for the first 3 seconds of her fall and again for times between $t = 0$ (when she jumps) and $t = 120$ seconds (when she lands).

3. Find an approximate value for her maximum velocity.
   HINT: Click in the plot window on the maximum and read off the t and v coordinates (at the left end of the toolbar).

4. Find an approximate value for her terminal velocity.

5. Find an exact value for her asymptotic velocity by computing $\lim\limits_{t \to \infty} v$.
   NOTE: You should be able to verify this limit by algebraically manipulating the formula for v.

6. Her landing will be safe if her velocity is less than 5 meters per second. Was her landing safe?

12.5 The Ant and the Blade of Grass

GOAL: In this project, you will use D to differentiate a function, construct a tangent line at a specific point and at a general point, plot a series of points connected by line segments, plot several things on the same graph, and solve an equation using fsolve.

BACKGROUND: Before you start, try each of the following examples.

- Differentiating functions.
  If a function f is defined using an arrow definition, for example
  ```
  >   f := x -> x^3;
  ```
  then its derivative is computed using Maple's D operator:
  ```
  >   D(f);
  ```
  Notice that the output is an arrow-defined function but it doesn't have a name. If you wish to give it a name, say Df, then you type
  ```
  >   Df:=D(f);
  ```

- Tangent lines.
  The line tangent to the curve $y = f(x)$ at the point $(a, f(a))$, is $y = f_{\tan}(x) = f(a) + f'(a)(x - a)$. For the function $f(x) = x^3$ discussed above, the tangent line at $x = 2$ can be defined as a function using:
  ```
  >   a:=2;
  >   ftan := x -> f(a) + Df(a)*(x-a);
  ```
  The equation of the tangent line is then
  ```
  >   y = ftan(x);
  ```
  HINT: The benefit of defining a:=2; instead of just typing 2 each time, is that if you need to compute the tangent line at several points, you can cut and paste the whole sequence of commands and only change the value of a before re-executing.

- Plotting several functions.
  The function $y = f(x)$ and its tangent line $y = f_{\tan}(x)$, defined above, can be plotted together on the interval $-5 \leq x \leq 5$ by enclosing the two functions in square brackets [ ] and using the Maple plot command.
  ```
  >   plot([f(x),ftan(x)], x=-5..5);
  ```

- Lists and plotting points.
  In Maple an ordered list is denoted by separating the items by commas and enclosing the list in square brackets [ ]. Thus, the point $(2, 5)$ is entered as [2,5], while a list of points might be
  ```
  >   octagon:=[ [1,0], [4,0], [5,1], [5,4], [4,5], [1,5],
  >   [0,4], [0,1], [1,0] ];
  ```
  If you plot this list of points and connect the dots, you should get an octagon.

```
> plot(octagon);
```

If you have several lists of points, enclose them in square brackets [ ].
Try the following.
```
> line1:=[ [1,3], [2,3] ]; line2:=[ [3,3], [4,3] ];
> you:=[ [1,2], [2,1], [3,1], [4,2] ];
> plot([octagon, line1, line2, you]);
```
What did you get?

Finally you can even plot lists of points together with functions and also specify the x-range, y-range and colors. For example
```
> plot([f(x), ftan(x), octagon, you], x=0..5, y=0..6,
> color=[red,blue,green,magenta]);
```

- Solving equations.

   Suppose you want to solve the equation $x/\pi + \sin(x) = 1$. You first enter the equation to make sure you have typed it properly.
```
> eq := x/Pi + sin(x) = 1;
```
You can then plot the left hand and right hand sides of this equation, to see where they intersect. Notice the use of the commands lhs and rhs.
```
> plot([lhs(eq),rhs(eq)], x=-2*Pi..4*Pi);
```
So there are three intersection points, i.e., solutions, one between 0 and 2, one between 2 and 4, and one between 4 and 6. To find the solutions, use Maple's fsolve command.
```
> fsolve(eq,x);
```
This finds the solution between 2 and 4. (It's π.) The other two can be found by adding ranges to fsolve:
```
> fsolve(eq,x=0..2);   fsolve(eq,x=4..6);
```
We now have all three solutions.

ASSIGNMENT: An ant is walking (to the right) over its ant mound, whose height (in cm) is given by the function: $h(x) = \dfrac{x^2 - 24x + 153928}{5(x^2 - 36x + 396)^2}$. Nearby there is a blade of grass, which is located as the line segment from $(40, 0.1)$ to $(40, 12)$. The goal in this lab is to find the point where the ant first sees the blade of grass. You can assume that the ant's line of sight is the tangent line to the ant mound. The following series of questions will lead you to the solution.

1. Define the function h that gives the height of the ant mound, using an arrow definition. Define the line segment occupied by the blade of grass and name it grass. Plot the ant mound in brown and the blade of grass in green on the same graph. Include the plot option, scaling=constrained.

2. Compute the derivative of h and name it Dh.

12.5. THE ANT AND THE BLADE OF GRASS

3. Compute the tangent line to $y = h(x)$ at $x = 16$ and define it as a function `htan` using an arrow definition. Plot the ant mound, the blade of grass, and the ant's line of sight when the ant is at $x = 16$. Can it see the blade of grass? Find the height H where the tangent line crosses the line $x = 40$ by evaluating `htan(40.)`.

4. Compute the tangent line to $y = h(x)$ at $x = 17.5$ and define it as a function `htan` using an arrow definition. Plot the ant mound, the blade of grass, and the ant's line of sight when the ant is at $x = 17.5$. Can it see the blade of grass? Find the height H where the tangent line crosses the line $x = 40$ by evaluating `htan(40.)`.

5. We can now see that, when the ant is at some position $x = a$ between 16 and 17.5, it can first see the top of the blade of grass. We want to find a. So, compute the tangent line to $y = h(x)$ at $x = a$ for a variable a. Define the tangent line at $x = a$ as a function `htan` using an arrow definition. NOTE: If you previously gave a a value, clear it by executing `a:='a';`

6. You can no longer plot the tangent line because its formula contains a variable, namely a. However, you can still find the height H where the tangent line crosses the line $x = 40$ by evaluating `htan(40.)`. When this height H equals the height of the blade of grass, the ant can just begin to see the blade of grass. Use Maple's `fsolve` command to solve for the value of a where H equals the height of the blade of grass. (You may need to specify a range for a in the `fsolve` command.) Denote the solution by A.

7. For the value A found in problem 6, compute the tangent line to $y = h(x)$ at $x = A$ and define it as a function `htan` using an arrow definition. Plot the ant mound, the blade of grass, and the ant's line of sight when the ant is at $x = A$. Can it see the blade of grass? Find the height H where the tangent line crosses the line $x = 40$ by evaluating `htan(40.)`.

8. There is a second solution to the equation $H = 12$. What is wrong with this solution?

9. *10% extra credit.* Animate the tangent line as the ant walks over the mound from $a = 7$ to $a = 19$. HINT: Use the `animate` command in the `plots` package.

12.6 Distance Between Two Curves

BACKGROUND: You need to know how to find the slope of the line between two points and the slope of a tangent line. You also need to know the condition on slopes which says that two lines are perpendicular to each other.

ASSIGNMENT: Given two curves that do not intersect, you want to find the shortest distance between the two curves. It is geometrically obvious (and can be justified mathematically) that the closest the curves come to each other are at the points P on the first curve and Q on the second curve such that the line from P to Q is perpendicular to both curves. More precisely, the line from P to Q is perpendicular to each of the tangnet lines at P and Q.

1. Start by considering the two curves $y = x^2$ and $y = \ln(x)$. Plot both curves on the same axes. Be sure to set scaling=constrained so that distances will be correct.

2. Let L_1 be the tangent line to $y = x^2$ at the point $P = (p, p^2)$.
   Let L_2 be the tangent line to $y = \ln(x)$ at the point $Q = (q, \ln(q))$.
   Let L_3 be the line between the points $P = (p, p^2)$ and $Q = (q, \ln(q))$.
   Find the slope m_1 of L_1, the slope m_2 of L_2, and the slope m_3 of L_3.

3. Write the equations which say L_3 is perpendicular to both L_1 and L_2. Solve these equations for p and q. Find P and Q.

4. Find the distance between the points P and Q.

5. Find the equations of the lines L_1, L_2 and L_3.
   Add the three lines to your plot.

6. Repeat steps 1-5 for each of the following pairs of curves:

   (a) $y = 2 + x^2$ and $y = 2x$
   (b) $y = e^x$ and $y = 2\ln(x)$
   (c) $y = \dfrac{1}{1+x^2}$ and $y = x^2 + \dfrac{1}{x^2}$ in the first quadrant

7. (Optional) Write a Maple procedure distbetwcrvs whose arguments are two functions and plotting ranges for x and y, and which prints out the closest points P and Q, plots the functions, tangent lines and the line between P and Q and returns the shortest distance between the two functions. If you write a procedure, you may use it to do Part 6.

12.7 Tangent and Normal Lines

BACKGROUND: Know how to compute the tangent and normal lines to a curve. Recall that he tangent line at $x = a$ has slope $f'(a)$ and the normal line has slope $-1/f'(a)$.

ASSIGNMENT: Consider the tangent and normal lines to the curve $y = x^2$.

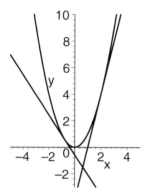 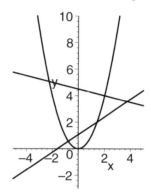

1. Find all points (p, q) which belong to exactly 2 distinct tangent lines to the curve $y = x^2$.
   HINT: Write the equation for the tangent line to $y = x^2$ at $x = a$. Write the equation which says (p, q) is on this line. This should be a quadratic equation to solve for a. When are there exactly 2 real solutions?

2. Find all points (p, q) which belong to exactly 2 (but not 3) distinct normal lines to $y = x^2$. Plot these points. This question is considerably harder than the first.
   HINT: Write the equation for the normal line to $y = x^2$ at $x = a$. Write the equation which says (p, q) is on this line. This can be rewritten as a cubic equation to solve for a. When are there exactly 2 real solutions? (A cubic equation $g(a) = 0$ has exactly 2 real roots only when there is a double root. This occurs when the double root is also a root of the derivative. So $g(a) = 0$ and $g'(a) = 0$ must both have a root at the same point.)

12.8 Parameterizing Letters

BACKGROUND: See Section 3.2 for a discussion of parametric curves. The goal of this project is to use parametric curves to form letters. Letters can be made with a combination of straight line segments, circles and ellipses which may be parameterized as follows:

- The line segment from the point (x_1, y_1) to the point (x_2, y_2) is parameterized by $x = x_1 + t(x_2 - x_1)$ and $y = y_1 + t(y_2 - y_1)$ for $0 \leq t \leq 1$. For example, the following Maple commands plot the line segment from $(1, 0)$ and $(2, 3)$:
    ```
    > x1,y1:=1,0; x2,y2:=2,3;
    > xt:=x1+t*(x2-x1); yt:=y1+t*(y2-y1);
    > plot([xt,yt,t=0..1], 0..3, 0..3, scaling=constrained);
    ```

- An arc of a circle centered at (a, b) of radius r can be parameterized by $x = a + r\cos(t)$ and $y = b + r\sin(t)$ for t in an appropriate range of angles. For example, the following Maple commands plot a three-quarter circle of radius 2 centered at $(-1, 3)$.
    ```
    > xt:=-1+2*cos(t); yt:= 3+2*sin(t);
    > plot([xt,yt,t=Pi/4..7*Pi/4], -4..2, -1..5,
    > scaling=constrained);
    ```

- Similarly, an arc of an ellipse, centered at the point (a, b) with horizontal radius r and vertical radius s is parameterized by $x = a + r\cos(t)$ and $y = b + s\sin(t)$.

- For example, the following commands plot the letter P:
    ```
    > P1:=plot([0,2*t, t=0..1]):
    > P2:=plot([0+.5*cos(t), 1.5+.5*sin(t), t=-Pi/2..Pi/2]):
    > with(plots):
    > display(P1,P2, scaling=constrained, axes=none);
    ```

ASSIGNMENT:

1. Plot the capital letters R and S, using straight line segments and arcs of circles or ellipses.

2. Repeat Exercise 1, but with the letters moved .5 units above the x axis and .5 units to the right of the y axis, and with its size doubled.

3. Use parametric curves to make graphs of the letters in your name.

12.9 Parametric Curves

GOAL: In this lab[1] you will use Maple to plot parametric curves, to find intersections of parametric curves with various lines, to find slopes of parametric curves and to find self-intersections of parametric curves.

BACKGROUND: Parametric curves and their Maple plots were discussed in Section 3.2. Recall from Section 2.3 that **fsolve** is best used in conjunction with a plot. If you're trying to find where a function equals zero, you should first plot it to make sure that **fsolve** is giving you all the roots, and to find the ranges to put into **fsolve**.

ASSIGNMENT: Consider the parametric curve

$$x = f(t) = 2\sin(2\pi t) - 2\cos^5(2\pi t) \qquad y = g(t) = \cos(2\pi t) - 3\sin(2\pi t)$$

for $0 \leq t \leq 1$. As you answer each question, be sure to refer back to the original plot to check that your answers make sense.

NOTE: "Find the points" with some property, means find the x and y coordinates.

1. Plot the curve with `scaling=constrained`. Find the points on the curve where t is $0, 0.1, 0.2, 0.3, 0.4, 0.5, 0.6, 0.7, 0.8$ and 0.9. Plot these 10 points and combine them with the plot of the curve. See Section 3.1 for point plots and combining plots. By hand (on a printed plot) or using `textplot`, label the points with t values.

2. Find the points where the curve crosses the line $x = 1$.
   HINT: You need to find the values of t which make the x coordinate equal to 1 by using **fsolve** and then use these t values to find the y coordinates. A plot of $f(t)$ as function a of t, may be helpful.

3. Find the points where the curve crosses the line $y = 2x$.
   HINT: Find the t values which solve $g(t) = 2f(t)$.

4. Find the points where the tangent line to the curve is horizontal.

5. Find the points where the tangent line to the curve is vertical.

6. Find the points where the tangent line has slope 3.
   HINT: This is the equation $\dfrac{dy}{dt} = 3\dfrac{dx}{dt}$.

7. Find the points where the curve crosses itself.
   HINT: You need to find two different values, $t = r$ and $t = s$, so that $f(r) = f(s)$ and $g(r) = g(s)$. Do this by using **fsolve** on this pair of equations, with ranges for r and s found from your plot of the curve. See the `?fsolve` examples.

---

[1] This project was originally developed by Thomas Vogel, Texas A&M Univ.

12.10 Seeing a Blimp

ASSIGNMENT: A blimp[2] has the shape and location given by the ellipse

$$\left(\frac{x - 1000}{96}\right)^2 + \left(\frac{y - 1200}{30}\right)^2 = 1$$

which may be parametrized by

$$x = 1000 + 96\cos(\theta), \qquad y = 1200 + 30\sin(\theta).$$

What is the angular size in degrees of the blimp as seen from your eye which is located at the origin, $O = (0,0)$. In other words, find two points, $P = (a,b)$ and $Q = (c,d)$, on the blimp, so that the angle between the vectors $\overrightarrow{OP}$ and $\overrightarrow{OQ}$ is as large as possible. Plot the ellipse and the line segments $\overline{OP}$ and $\overline{OQ}$.

HINTS:

- The `angle` command in the `linalg` package can be used to find the angle in radians between two vectors. Only convert the final angle to degrees.

- Since the ellipse is longer in the x-direction, you would expect the points P and Q to be close to the endpoints of the x-diameter; i.e. close to the points $P_o = (904, 1200)$ and $Q_o = (1096, 1200)$. Find the angle between the vectors $\overrightarrow{OP_o}$ and $\overrightarrow{OQ_o}$. The actual angle should be slightly larger than this.

- If you do nothing to restrict the `fsolve` command, Maple is most likely to miss the maximum and find a minimum of 0 angle which occurs whenever $P = Q$. So in the `fsolve` command, restrict the ranges of the variables so that P is close to P_o and Q is close to Q_o.

- To solve the problem, first hold $Q = Q_o$ and vary $P = (a,b)$. Then hold $P = P_o$ and vary $Q = (c,d)$.

- If you solve the problem by eliminating a variables, express a in terms of b in the half of the ellipse near P_o and express c in terms of d in the half near Q_o.

- If you solve the problem by parametrizing the ellipse, then P is close to P_o when θ is close to π and Q is close to Q_o when θ is close to 0.

---

[2] The dimensions and altitude of the Goodyear blimp were taken from http://www.goodyearblimp.com/q_construct.html#sizes and http://www.goodyearblimp.com/q_flying.html#high

12.11 Static and Dynamic Tension

BACKGROUND: Newton's Second Law says $\sum \vec{F} = m\vec{a}$, i.e. the sum of the vector forces on an object is equal to its mass times its acceleration vector. The situation is "static" when the object is not moving and $\vec{a} = \vec{0}$. It is "dynamic" when the object is moving and $\vec{a} \neq \vec{0}$.

ASSIGNMENT: A weight of mass $m = 10$ kg is supported at point $C = (x, y)$ by two stiff rods supported at points $A = (0,0)$ and $B = (z,0)$. Point A is held fixed while point B is free to slide horizontally along a rail. The rod AC has length $L_a = 3$ m. The rod BC has length $L_b = 5$ m. The force of gravity is $\vec{F}_g = (0, -mg)$ where $g = 9.8$ m/sec^2. The tension forces along the rods AC and BC are $\vec{T}_a = (-T_a \cos \alpha, T_a \sin \alpha)$ and $\vec{T}_b = (T_b \cos \beta, T_b \sin \beta)$ respectively.

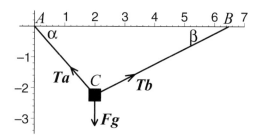

1. For what value of z is $\alpha = \pi/2$? What are the smallest and largest values of z? (NOTE: The angle α can be larger than $\pi/2$.) No Maple required.

2. The position of $B = (z, 0)$ determines α, β and the position of $C = (x, y)$. Set up and solve two equations for α and β as functions of z. Find parametric equations for x and y as functions of the parameter z. Plot x and y. Discuss the plots.

3. If the point $B = (z, 0)$ is held fixed, the situation is static. Set the sum of the forces equal to zero and solve for T_a and T_b, as functions of z. Plot T_a and T_b. What happens to T_a and T_b when $z = 4$? When $z < 4$? Give physical explanations.

4. When $B = (z, 0)$ moves, $C = (x, y)$ also moves. So the situation is dynamic. Assume $B = (z, 0)$ starts at $(2, 0)$ at $t = 0$ and moves to the right with constant velocity with $\dfrac{dz}{dt} = 1$ m/sec. Find $x(t)$, $\dfrac{dx}{dt}$, $\dfrac{d^2 x}{dt^2}$, $y(t)$, $\dfrac{dy}{dt}$ and $\dfrac{d^2 y}{dt^2}$. Set the sum of the forces equal to the mass times the acceleration and solve for T_a and T_b as functions of t. Then convert to functions of z. Plot T_a and T_b as functions of z. Discuss the acceleration and tension near $t = 2$, $(z = 4)$.

5. Compare the static and dynamic plots for T_a and T_b. Which are larger?

12.12 Calculus I Review

BACKGROUND: In this lab, you will review the domain of a function, composition of functions, differentiation (using the chain rule), and integration (to compute area), as well as the Maple commands for these operations.

ASSIGNMENT: In problems 1-9, consider the following four functions:

$$f(x) = x^2, \quad g(x) = \sqrt{4-x}, \quad h(x) = (f \circ g)(x), \quad k(x) = (g \circ f)(x)$$

1. Enter f and g into Maple using arrow definitions, e.g.

    > f:=x->x^2;

2. Enter h and k into Maple using Maple's composition symbol, @. Compute Maple's expressions for $h(x)$ and $k(x)$, e.g.

    > h:=f@g;   h(x);

3. Using Maple's plot command, graph f and g for $-6 \le x \le 6$ in one plot. Make f red and thin. Make g blue and thick.

4. The function f has all real numbers for its domain. What is the domain for the function g? Explain why the plot for g is not drawn for all numbers between -6 and 6.

5. Graph h and k for $-6 \le x \le 6$ in one plot. Make h red and thin. Make k blue and thick.

6. Explain why the straight line $y = 4 - x$ should not be the graph of h, and draw a better graph.

7. Compute the derivatives of $h(x)$ and $k(x)$. Are Maple's answers correct? Why?
    NOTE: You can have Maple calculate derivatives with either the D operator or the diff operator.

    > deriv_h:=D(h)(x);
    > deriv_h:=diff(h(x),x);

8. Sketch the region in the first quadrant bounded by the curves $y = 0$, $y = f(x)$ and $y = g(x)$.

9. Calculate the area of the region you sketched in problem 8, by using the Maple's Int and value commands
    NOTE: Instead of using Maple's Int and value commands, you could use int. If you do this Maple will not print out the integrals; only the answer will be printed. *Get in the habit of using commands that print out what you want calculated.* This makes it a lot easier to find and correct typos and other mistakes.

10. Explain the differences in output between the following approaches to calculating the derivative of the function $\dfrac{1}{x(x+2)}$.

(a) > diff(1/x*(x+2),x);
(b) > p:=1/x*(x+2); diff(p,x);
(c) > p:=(x*(x+2))^(-1); diff(p,x);
(d) > Diff(1/x*(x+2),x); value(%);

Which values are correct? How can the others be fixed? After fixing them, which approach do you feel is best? Why?

12.13 Calibrating a Dipstick

BACKGROUND: Review volumes of revolution.

ASSIGNMENT:

1. Suppose the quarter-circle $y = -\sqrt{64 - x^2}$ for $0 \leq x \leq 8$ is rotated about the y-axis to create a hemispherical bowl (8 units deep). Construct a dipstick which when inserted vertically into the bowl will determine when the bowl is one quarter full, one-half full and three-quarters full. Plot the percent full as a function of the height on the dipstick. Then plot the height on the dipstick as a function of the percent full.

   HINT: Recall the discussion of plotting inverse functions in Section 3.3.

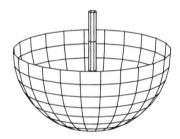

2. Now repeat the same problem assuming that the bowl is generated by rotating the curve $y = -(256 - x^4)^{1/4}$ for $0 \leq x \leq 4$ around the y-axis.

12.14 The Area of a Unit p-Ball

GOAL: In this project, you will determine the area of a unit p-ball in the plane for different values of p and look at their limiting characteristics.

DEFINITIONS: The p-norm of a vector $\vec{v} = (x, y)$ is $|\vec{v}|_p = \sqrt[p]{|x|^p + |y|^p}$ instead of the standard Euclidean 2-norm $|\vec{v}|_2 = \sqrt{|x|^2 + |y|^2}$. So a p-normed circle of radius R with center at the origin is the set of points (x, y) satisfying

$$|x|^p + |y|^p = R^p,$$

and a p-ball is the interior of a p-normed circle. So you need to compute the area of the region satisfying $\quad |x|^p + |y|^p \leq 1$.

1. Using implicitplot or just plot with scaling=constrained, graph several unit p-circles in the plane with $p \geq 1$. Specifically, superimpose the curves $|x|^p + |y|^p = 1$ for $p = 1, 2, 3, 4, 5$. Notice they are convex.

2. Make a conjecture as to the limiting shape and area of these p-balls as $p \to \infty$.

3. Using implicitplot or just plot with scaling=constrained, graph several unit p-circles in the plane with $0 < p \leq 1$. Specifically, superimpose the curves $|x|^p + |y|^p = 1$ for $p = 1, \frac{1}{2}, \frac{1}{3}, \frac{1}{4}, \frac{1}{5}$. Notice they are concave for $p < 1$.

4. Make a conjecture as to the limiting "shape" and "area" of these p-balls as $p \to 0^+$.

5. For $p = 1, 2, 3, 4, 5$, compute the area of the unit p-ball $|x|^p + |y|^p \leq 1$.

    HINT: For each value of p, the fact that the p-ball is symmetric with respect to both the x-axis and the y-axis means that the total area is 4 times the area of the part of the p-ball in the first quadrant.

6. Obtain a general formula for the area of the unit p-ball for $p \geq 1$.

7. What is the limiting value of the area of the unit p-ball as $p \to \infty$? Apply Maple's Limit and value commands to either the integral or your general formula. Does this agree with your conjecture?

8. For $p = \frac{1}{2}, \frac{1}{3}, \frac{1}{4}, \frac{1}{5}$, compute the area of the unit p-ball $|x|^p + |y|^p \leq 1$.

9. Obtain a general formula for the area of the unit p-ball for $0 < p < 1$.

    HINTS: Let $p = \dfrac{1}{q}$. The formulas in #6 and #9 are the same.

10. What is the limiting value of the area of the unit p-ball as $p \to 0^+$? Does this agree with your conjecture?

12.15 The Center of the State of Texas

BACKGROUND: Review center of mass and numerical integration. Recall that the center of mass of a region bounded above by $y = f(x)$ and below by $y = g(x)$ for $a \leq x \leq b$ is the point (x_0, y_0) where

$$x_0 = \frac{1}{A} \int_a^b x\,[f(x) - g(x)]\,dx \quad \text{and} \quad y_0 = \frac{1}{A} \int_a^b \frac{1}{2} \left([f(x)]^2 - [g(x)]^2\right) dx$$

and A represents the area of the region. In the problem given below, there are no analytical formulas for f and g, just numerical data. So the integrals must be computed numerically.

ASSIGNMENT: The goal of this project is to approximate the center of mass of the state of Texas from data that represent the state's boundary. The northern and southern boundaries are given by

```
> north:=[[0,0], [1,0], [2,0], [3,0], [3,4.5], [4,4.5],
> [5,4.5], [6,4.5],[6,2.2], [7,2.1], [8,1.8], [9,1.9],
> [10,1.8], [11,1.7], [11,-2.2]];
> south:=[[0,0], [1,-1.1], [2,-2.5], [3,-2.9], [4,-2.3],
> [5,-2.8], [6,-4.4], [7,-5.8], [8,-6.1], [9,-3.3],
> [10,-2.8], [11,-2.2]];
```

Here, the origin is the western corner of Texas (near El Paso) and the x-axis is the extension of the east-west border between New Mexico and Texas. Each unit represents approximately 69 miles.

NOTE: There are two y-values given in the northern boundary for both $x = 3$ and $x = 6$ (because $x = 3$ and $x = 6$ represent the two north-south boundaries of the Panhandle). Use these data to answer these questions.

1. Plot Texas. (See Exercise 2 of Chapter 3.)

2. Compute the area of Texas using the trapezoid rule to compute the integrals. (See Exercises 10 and 11 of Chapter 6.)

3. Compute the center of mass of Texas using the trapezoid rule.
   NOTE: In the formulas for center of mass, the function f is approximated by (the y-components in) the north data, while g is approximated by the south data,

4. Plot Texas and its center of mass. (See Section 3.1 to see how to combine plots using display.)

12.16 Center of Gravity of a Parabolic Plate

BACKGROUND: Know how to compute the normal line to a curve and the center of mass of a region. Recall that the normal line to the curve $y = f(x)$ at the point $(c, f(c))$ has slope $m = \dfrac{-1}{f'(c)}$. Also the center of mass of a region bounded above by $y = g(x)$ and below by $y = f(x)$ for $a \leq x \leq b$ is the point (x_0, y_0) where

$$x_0 = \frac{1}{A} \int_a^b x\, [g(x) - f(x)]\, dx \quad \text{and} \quad y_0 = \frac{1}{A} \int_a^b \frac{1}{2} \left([g(x)]^2 - [f(x)]^2\right) dx$$

and A represents the area of the region.

ASSIGNMENT: Consider a thin metal plate with uniform density which occupies the region between the parabola $y = \dfrac{1}{30}x^2$ and the line $y = 120$ where the y-axis is vertical and the x-axis is the ground. In this position the center of gravity is located on the y-axis at $(0, 72)$. If the plate is rolled slightly along the x-axis, the plate will become "off-balance" and will roll to a new equilibrium state.

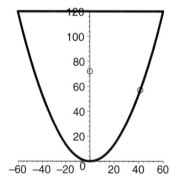 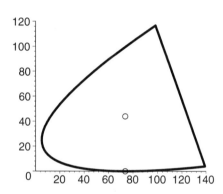

1. Verify that the original center of mass is at $(0, 72)$.

2. Give a geometric and physical argument which explains why the plate will not return to its original position and where it will roll to.

3. Find the point on the edge of the plate in the original position which will be touching the ground in the new equilibrium position.

4. Give the coordinates of the center of gravity of the plate in its new equilibrium position. Assume it rolls without slipping.

5. Repeat problems 3 and 4 for the generic parabola $y = ax^2$ where $a > 0$ (still use the line $y = 120$).
   HINT: Start by executing `assume(a>0);`.

6. Give a parametric plot (with the parameter a) of the centers of gravity found in problem 5. What's the smallest allowed value of a?

12.17 The Oil Tank

BACKGROUND: Review the following applications of integration:
  How to compute a volume by slicing.
  How to find the work done to pump a liquid out of a tank.
  How to find the center of mass of a solid.

ASSIGNMENT: The oil tank[3] below is cylindrical with hemispherical "caps". The length of the cylinder part is $L = 40\,ft$ and the radius of the cylinder (and of each hemispherical cap) is $r = 5\,ft$.

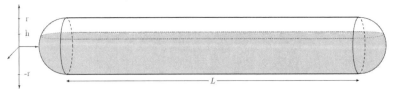

The tank contains $V = 22,300\,gal$ of Texas crude oil whose density is $\rho = 54.5\,lb/ft^3$. Find:

1. The height h of the oil above the center of the tank.

2. The weight (or force induced by gravity) of the oil.

3. The amount of work required to pump the oil out of the top of the tank.

4. The distance below the center of the tank where the center of gravity of the oil occurs.

You may assume that the acceleration of gravity is $g = 32\,ft/sec^2$ and $1\,ft^3 = 7.481\,gal$.

---

[3] This project was originally developed by Troy Henderson, Texas A&M Univ.

12.18 The Skimpy Donut

GOAL: In this project, you will compute a volume of revolution, a surface area of revolution, and solve a max/min problem.

ASSIGNMENT: The GETFAT Donut company makes donuts with a thin layer of chocolate icing. The company decides to cut costs by minimizing the amount of chocolate icing used on each donut without shrinking the volume of the donut.

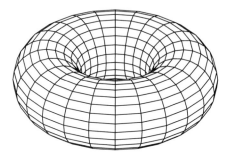

The donut has the shape of a torus which is formed by revolving the circle $(x-a)^2 + y^2 = b^2$ around the y-axis. (Here b is the radius of the circle and a is the distance from the center of the hole to the center of the circle.) At present the company makes donuts with $a = 5$ cm and $b = 3$ cm. The problem, then, is to determine the dimensions of the donut with the same volume that minimize the surface area. Follow these steps:

1. Compute the volume V of the donut as a function of a and b using the technique of either washers or cylindrical shells. As a check, verify the volume of the donut with $a = 5$ and $b = 3$ is $90\pi^2 \approx 888$ cm^3.
   HINT: First execute `assume(b>0, a>b);`

2. Compute the surface area of the donut as a function of a and b either by solving the equation of the circle for x or y or by parametrizing the circle as $x = a + b\cos t$, $y = b\sin t$. As a check, verify the surface area of the donut with $a = 5$ and $b = 3$ is $60\pi^2 \approx 592$ cm^2.

3. With the volume fixed at $90\pi^2$, find the dimensions a and b of the donut which minimize the surface area of the donut.
   NOTE: For this problem, you will have to determine the range of allowable a and b.

4. Is there a maximum surface area for the given volume? (So the company could advertise extra chocolate.)

5. Write a report to the CEO summarizing your recommendations (including the percent savings or percent extra cost). Anything you say in this report must be documented in an appendix of Maple computations for the engineers.

12.19 Area Between a Curve and Its Tangent Line

ASSIGNMENT: In this project,[4] you will find the tangent line to the graph of a function for which the area between the curve and the tangent line is a minimum.

1. Pick a function $y = f(x)$ which is everywhere concave up or everywhere concave down, such as $y = f(x) = -x^2$. NOTE: If the concavity changes, then the tangent line might cross the curve, which we don't want.

2. Find its tangent line at a general point $x = p$.

3. Compute the area between the curve and its tangent line at $x = p$ above the interval $0 \le x \le 1$. Label it Area.

4. Find the point $x = p_{\min}$ for which Area is a minimum. Be sure to apply the Second Derivative Test to verify that your critical point is a minimum.

5. Plot the curve and the tangent line for several values of p in $[0, 1]$ including the minimum. Plot the Area function.

6. Repeat steps 1-5 for three or more other functions $f(x)$. Use interesting functions, not just polynomials, and check the concavity on the interval $[0, 1]$. Be sure to try functions which are concave up as well as concave down.

7. What do you conjecture?

8. Prove your conjecture by repeating steps 1-4 for an undefined function f:=g(x), once assuming g is concave up and once assuming g is concave down. Before solving for p you will need to give names to the derivatives of g using, for example, subs(diff(g(p),p,p)=ddg, ...).

9. What happens to your conjecture and proof if you change the interval from $[0, 1]$ to $[a, b]$?

---

[4]The idea for this project was originally suggested by Carol Scheftic, Cal. Poly. St. Univ.

12.20 Curves Generated by Rolling Circles

BACKGROUND: Review parametric curves and their slope and arc length.
ASSIGNMENT: In this project, you will study the cycloid, the epicycloid and the hypocycloid, which are curves generated by a point on a rolling circle.

1. Consider a wheel of radius R. Fix a point P on the rim of the wheel, initially at the bottom. Now let the wheel roll on level ground and consider the path traced out by the point P. (See the figure.) This path is called a *cycloid*.

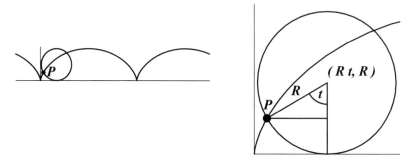

   (a) Show that the cycloid is parameterized by the formulas
   $$x(t) = R\left(t - \sin(t)\right) \quad \text{and} \quad y(t) = R\left(1 - \cos(t)\right)$$
   Here, θ is the angle between the vertical and the ray that extends from the center of the circle to P (so $\theta = 0$ when P is at the origin). Enter them into Maple as: `x1,y1:=R*(t-sin(t)),R*(1-cos(t));`

   (b) Plot two arches of this cycloid with $R = 1$. To plot a parametric curve (x_1, y_1) (where x_1 and y_1 are expressions in t) over the interval $a \le t \le b$, use the command: `plot([x1,y1,t=a..b]);`
   NOTE: You must use square brackets with this plot command. Now unassign R using `R := 'R';`

   (c) The arc length of a parametric curve $(x(t), y(t))$ for $a \le t \le b$ is
   $$\int_a^b \sqrt{\left(\frac{dx}{dt}\right)^2 + \left(\frac{dx}{dt}\right)^2}\, dt$$
   Compute the arc length of one arch of this cycloid for a general value of R. (If necessary, `assume` R is positive. See `?assume`.)

   (d) The slope of a parametric curve $(x(t), y(t))$ is given by $\dfrac{dy}{dx} = \dfrac{dy/dt}{dx/dt}$. Find the slope of the cycloid at $t = \dfrac{\pi}{3}$.

   (e) Find the slope of the cycloid as $t \to 0^+$ by computing an appropriate limit. (See `?limit[dir]` to learn about one-sided limits.)

2. Now suppose a circle of radius a rolls around the outside of a circle of radius $R > a$ centered at the origin. (See the figure.) The path of a fixed point P on the rolling circle is called an *epicycloid*.

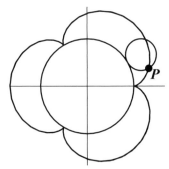
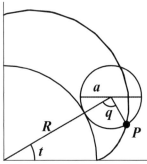

(a) Find the parameterization the epicycloid. Here t is the angle measured counterclockwise from the positive x-axis to the line segment that runs from the origin to the center of the rolling circle. Assume that P is located at the point $(R, 0)$ when $t = 0$.
HINT: Show that the angle q in the figure is $q = Rt/a$.

(b) Plot the epicycloid with $R = 3$ and $a = 1$.

(c) Compute the arc length of one of the arches of the epicycloid. (Assume R is positive, a is positive and $R > a$.)

3. Now suppose a circle of radius a rolls around the inside of a circle of radius $R > a$ centered at the origin. (See the figure.) The path of a fixed point P on the rolling circle is called an *hypocycloid*.

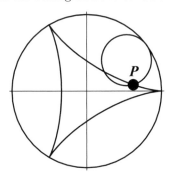
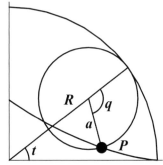

(a) Find the parameterization the hypocycloid. Here t is the angle measured counterclockwise from the positive x-axis to the line segment that runs from the origin to the center of the rolling circle. Assume that P is located at the point $(R, 0)$ when $t = 0$.
HINT: Show that the angle q in the figure is $q = Rt/a$.

(b) Plot the hypocycloid with $R = 3$ and $a = 1$.

(c) Compute the arc length of one of the arches of the hypocycloid. (Assume R is positive, a is positive and $R > a$.)

12.21 The Wankel Rotary Engine

BACKGROUND: You need to know how to plot polygons and parametric curves, how to compute the area between two curves and how to use the `piecewise` and `animate` commands. In particular the area under a parametric curve $(x(t), y(t))$ is

$$A = \int_{x_1}^{x_2} y(x)\, dx = \int_{t_1}^{t_2} y(t) \frac{dx}{dt}\, dt$$

Further the length of an arc of a circle with central angle θ is $s = r\theta$.

GOAL: In this project you will study the motion of Felix Wankel's rotary engine[5] of the type once used by Mazda. The basic geometry consists of two circles and an equilateral triangle which just fits inside a curve called an epitrochoid.

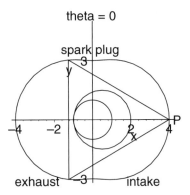

In our simple model, the inner circle has radius 1, center at the origin and does not move. The outer circle has radius $3/2$, is always tangent to the inner circle and is constrained to roll counterclockwise without slippage around the inner circle. The equilateral triangle is concentric with the outer circle and has vertices which are at distance $7/2$ from the center of the triangle. This triangle, called the rotor, rolls with the outer circle. At time $t = 0$ the two circles are tangent to each other at the point $x = -1$ while one vertex of the triangle (called P) is on the positive x-axis at $x = 4$ as shown above. As time progresses, the outer circle rolls around the inner circle and the vertex P traces out a parametric curve $P(\theta) = (x(\theta), y(\theta))$ as a function of the angle θ through which the point of tangency has moved on the inner circle, measured counterclockwise from the negative x-axis. This curve is called an epitrochoid, and is the shape of a cross section of the "cylinder" of a Wankel engine.

In the engine, each of the three spaces between the rotor and the walls of the cylinder is called a "chamber". As the rotor revolves, each chamber passes through the various phases of the Otto cycle: injection, compression, ignition, expansion and exhaust. We will describe these phases for the chamber in front of the vertex P:

---

[5]For more information, see the Wikipedia entry on the Wankel engine.

12.21. THE WANKEL ROTARY ENGINE

From $\theta = -5\pi/2$ to $\theta = -\pi$, the fuel-air mixture is injected through the open intake valve on the lower right side:

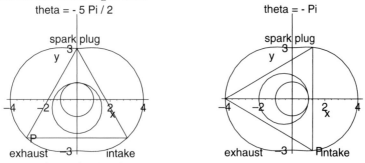

From $\theta = -\pi$ to $\theta = \pi/2$, the fuel-air mixture is compressed.
Just after maximum compression at $\theta = \pi/2$, the spark plug ignites the fuel.

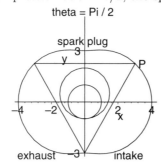

From $\theta = \pi/2$ to $\theta = 2\pi$, the fuel expands providing a power stroke.
From $\theta = 2\pi$ to $\theta = 7\pi/2 \approx -5\pi/2$, the used fuel is compressed and escapes through the open exhaust valve on the lower left side.

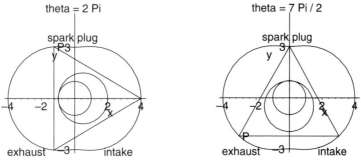

ASSIGNMENT:

1. The inner circle does not move. Write an equation, `eq1`, for the inner circle and use `implicitplot` to plot it.

2. The center of the outer circle (which is also the center of the triangular rotor) changes with the angle to the point of tangency with the inner circle, θ. Write an equation, `eq2`, for the outer circle for general θ. Then use `subs(theta=Pi/3,eq2)` and `implicitplot` to plot it when $\theta = \pi/3$.

3. Explain why, if the point of tangency of the two circles has moved counterclockwise around the small circle through θ radians, then the vertex P which was originally on the x-axis will be at

$$x_1 = \frac{1}{2}\cos(\theta) + \frac{7}{2}\cos\left(\frac{\theta}{3}\right) \quad \text{and} \quad y_1 = \frac{1}{2}\sin(\theta) + \frac{7}{2}\sin\left(\frac{\theta}{3}\right).$$

HINTS: The main two parts of the derivation are showing (i) that the center C of the triangle rotates through an angle θ and (ii) that the vertex P rotates through an angle $\theta/3$ around the center C. Below is a diagram which shows the inner and outer circles after the point of tangency has rotated by θ radians.

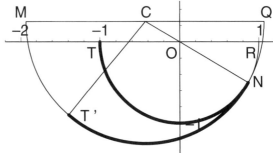

In the diagram, the center of the (small) stationary circle is at the origin, O. The center of the (large) rotor circle (and the triangle) is at C. The two circles are currently tangent at N. The points T and T' are the points on the inner and outer circles which used to be at the initial point of tangency and the thickened arcs represent the parts of the two circles that have been in contact with each other. Notice that N is diametrically opposed to C and that T' is diametrically opposed to the vertex P of the triangle (not shown). In other words, N, O and C are colinear and P, C and T' are colinear. So you need to show $\angle ROC = \theta$ and $\angle MCT' = \theta/3$. Given that $\angle TON = \theta$, identify each of the following and justify your answer: (When you answer (b) and (d), you will need to use the arclength formula $s = r\theta$ where θ is in radians, $r = 1$ for the small circle and $r = 3/2$ for the large circle.)

(a) the angle $\angle ROC$ (which completes part i)
(b) the arclength $\widehat{TN}$
(c) the arclength $\widehat{T'N}$
(d) the angle $\angle T'CN$
(e) the angle $\angle NOR$
(f) the angle $\angle NCQ$
(g) the angle $\angle MCT'$ (which completes part ii)
(h) the coordinates of the point C.
(i) the coordinates of the point P.

12.21. THE WANKEL ROTARY ENGINE

4. Use the previous result to do a parametric plot of the (epitrochoid) cylinder wall along which the vertex P of the triangular rotor travels.

5. The position of the vertex P of the rotor is given in #3 above. The other two vertices of the triangular rotor are $2\pi/3$ apart. Write down the positions of the other two rotors as (x_2, y_2) and (x_3, y_3) for general θ. Then use `subs` and `plot` to plot the triangle when $\theta = \pi/3$.

6. Combine the plots of both circles, the triangle and the epitrochoid for $\theta = \pi/3$ using `display`.

7. Write a Maple procedure `F(theta)` which plots both circles, the triangle and the epitrochoid at the same time as a function of θ. (See Section 10.3.)

8. Make the outer circle and the triangle roll around the inner circle by animating your plot. (See `?animate`.) To see it move, you need to right click in the plot and select ANIMATE > PLAY or click in the plot and then click on the PLAY button which is a triangle pointing to the right on the toolbar.

9. Write a Maple procedure with argument θ to compute the area between a side of the rotor and the epitrochoid. Observe that for many angles the top or bottom curve will be split into two parts: an edge of the rotor and a piece of the epitrochoid. Use Maple's `piecewise` command to make sure that the correct rules are applied at the correct angles. Look at your animation to determine the angles at which the changes occur.

10. Plot the graph of the area between the rotor and the wall over the interval $(0, 3\pi)$. From the plot, what are the values of θ (expressed as multiples of π) which give the minimum and maximum area? Using your area function, what are the minimum and maximum area?

11. The difference between the maximum and minimum area is known as the *displacement* of the engine. The ratio of the maximum area to the minimum area is called the *compression ratio*. Compute the displacement and compression ratio for your Wankel engine. (Higher compression ratios can be achieved by "rounding" the rotor in an appropriate manner, but we will not deal with that here.)

12.22 Shakespeare's Shylock

BACKGROUND: This lab introduces compound interest and continuous compounding. In the process, you will discover an important limit. You will need the Maple commands for functions of several variables (See Section 2.1.) and computing limits at infinity. (See Sections 5.3 and 9.1.)

ASSIGNMENT: Shakespeare's Shylock used to lend money at the usurious rate of 100% per year simple interest. Thus if you borrow $1000 (and don't pay any back) then at the end of 1 year you owe $2000, and at the end of 2 years you owe $3000 dollars. Being financially savy, Shylock realized he could make more money if he compounded the interest annually, or more frequently or even continuously. Let's investigate what he discovered. We first have some preliminary questions, and then we will come back to Shylock.

1. Suppose you put $1000 in a bank at 10% compounded annually for 2 years and don't make any withdrawls. (Use Maple as a calculator.)

    (a) How much interest will you receive at the end of the first year?

    (b) How much money will you have in the bank at the end of 1 year?

    (c) For the second year, the money you have at the beginning of the second year will receive 10% interest. How much interest will you receive at the end of the second year?

    (d) How much money will you have in the bank at the end of 2 years?

2. Suppose you put P (principal) in a bank which pays $100r$% interest compounded annually. (If the rate is 5%, then $r = .05$.)

    (a) In terms of P and r, how much will be the bank at the end of 1 year?

    (b) After 2 years?

    (c) After t years? Why?
    HINT: Factor your answer to part (b) using the factor command.
    ANSWER: $A = P(1+r)^t$
    Each year, the amount in the bank is multiplied by a factor of $(1+r)$. Multiplying the previous amount by the 1 says this money is still in the bank. Multiplying by the r gives the interest for the year. Adding these together gives the principal at the end of the year.

    (d) Express your answer as a Maple function A of the three variables P, r and t, as follows
       > A := (P,r,t) -> P*(1 + r)^t;

    (e) Use your Maple function A to compute the amount in the bank if $3000 is invested for 5 years at 6% interest compounded annually.

3. Suppose you put $1000 in a bank which pays 10% interest compounded semi-annually (i.e. twice a year you receive 5% interest). (Again, use Maple as a calculator.)

    (a) How much is in the bank after 6 months?

(b) After 1 year?

(c) After 2 years?

4. Suppose you put $P in a bank which pays $100r\%$ interest compounded n times a year (i.e. n times a year you receive $\dfrac{100r}{n}\%$ interest).

   (a) In terms of P, r and n, how much money will you have in the bank at the end of 1 year?
   (b) After 2 years?
   (c) After t years? Why?
   (d) Express your answer as a Maple function A of the four variables P, r, n and t.
   (e) Use your Maple function A to compute the amount in the bank if $3000 is invested for 5 years at 6% interest compounded monthly.

5. Now back to Shylock. Suppose you borrow $1000 at 100% interest ($r = 1.00$) compounded n times a year. How much money will you owe after 1 year, if you don't pay anything back and it is compounded

   (a) semi-annually? (Use your formula from 4(d).)
   (b) monthly?
   (c) weekly? (Assume 52 weeks in a year.)
   (d) daily? (Assume 365 days in a year.)
   HINT: At this point, execute `Digits := 20;` and be sure to put a decimal point in the rate: $r = 1.00$.
   (e) every minute?
   (f) every second?
   (g) continuously?
   HINT: Use Maple's `Limit` and `value` commands. You will be able to derive this limit once you learn l'Hospital's rule.

6. Write the answer to 5(g) in terms of known mathematical constants.

7. Now let's be realistic. Suppose you put $1000 in a bank at 5% interest compounded continuously.

   (a) How much would you have at the end of 1 year? (Use your formula from 4(d) and Maple's `Limit` and `value` commands.)
   (b) Compute the number $e^{.05}$ and compare it to your answer to (a). NOTE: In Maple, this is entered as `exp(.05)`.

8. Suppose you put $P in a bank at $100r\%$ interest compounded continuously.

   (a) How much money you will you have in the bank after 1 year? (Use your formula from 4(d) and Maple's `Limit` and `value` commands.)
   (b) How much money will you have in the bank after t years? Justify the answer.

12.23 The Bouncing Ball

BACKGROUND: Review geometric series and Newton's Laws.

ASSIGNMENT: A ball is dropped from a height of 54 feet. Each time it bounces it reaches a height which is 2/3 of the height on the previous bounce. Thus after the first bounce it reaches 36 feet, then 24 feet, then 16 feet, etc.

1. What is the total distance travelled by the ball (after an infinite number of bounces)? HINT: Use Sum and value.

2. What is the total time the ball takes to travel this distance? Assume Newton's Law of gravity with no air resistance. Thus, the distance h the ball falls is related to the time t by $h = \frac{1}{2}gt^2 = 16t^2$.
   HINT: You may need the rationalize command.

12.24 Pension Funds

BACKGROUND: Review separable differential equations and exponential growth.

ASSIGNMENT: A pension fund starts out with P (at $t = 0$) and is invested with a return of $100r\%$ per year, compounded continuously. (Here, r is the interest rate, given as a number between 0 and 1). The pension fund must continuously pay out money at the rate of R per year to its employees for a period of n years. (This means that the value of the pension fund decreases to zero after n years.) Let $y(t)$ denote the value of the pension fund after t years.

1. From the information given, derive the differential equation $\frac{dy}{dt} = ry - R$, and the conditions $y(0) = P$, $y(n) = 0$.

2. Solve this differential equation with the initial condition $y(0) = P$ for $y(t)$ by using dsolve.

3. Using the condition $y(n) = 0$, find a formula for P in terms of r, n and R. P represents the amount of money required to pay out R per year for n years, assuming the rate of return on the investment is $100r\%$.

4. Calculate P for $R = 40,000$, $r = .05$, and $n = 15$.

5. Calculate the interest rate ($100r\%$) required so that an initial value of $P = \$400,000$ for the pension fund will pay out $R = \$40,000$ per year for $n = 15$ years.

12.25 The Flight of a Baseball

BACKGROUND: Review motion in two dimensions, including velocity, acceleration and Newton's Laws.

ASSIGNMENT: Imagine that a baseball player is up at home plate and hits the ball into the outfield. What parametric equations describe the position of the ball t seconds after it is hit? How far will the ball travel? How fast does the ball need to be hit in order for the ball to clear the home run fence? This lab is designed to answer these questions.

1. First consider a simplified model that ignores air resistance. In this case, after the ball is hit, the only force acting on the ball is the vertical force due to gravity. Therefore, Newton's Laws say the x- and y-components of the acceleration of the ball are

$$\frac{d^2x}{dt^2} = 0 \quad \text{and} \quad \frac{d^2y}{dt^2} = -g$$

where $g = 32$ ft/sec^2 is the acceleration of gravity. Integrating these equations with respect to t gives

$$\frac{dx}{dt} = A \quad \text{and} \quad \frac{dy}{dt} = -gt + B$$

where A and B are constants of integration. Setting $t = 0$ in the above equations, says A and B are the x- and y-components of the initial velocity of the ball, respectively. Suppose the initial speed of the ball is v_0 (in units of feet per second) and the initial angle of inclination of the ball is θ. Then $A = v_0 \cos(\theta)$ and $B = v_0 \sin(\theta)$ and the equations become

$$\frac{dx}{dt} = v_0 \cos(\theta) \quad \text{and} \quad \frac{dy}{dt} = -gt + v_0 \sin(\theta)$$

One more integration with respect to t yields

$$x = v_0 t \cos(\theta) + C \quad \text{and} \quad y = \frac{-1}{2}gt^2 + v_0 t \sin(\theta) + D$$

where C and D are constants of integration. Setting $t = 0$ and assuming home plate is at origin ($x = 0, y = 0$) and the batter hits the ball at height h ft, then $C = 0$ and $D = h$. The final parameterization of the baseball's path is given by

$$x = v_0 t \cos(\theta) \quad \text{and} \quad y = \frac{-1}{2}gt^2 + v_0 t \sin(\theta) + h$$

Enter these formulas into Maple as expressions x1 and y1 with $g = 32$ ft/sec^2, $\theta = \pi/4$ rad, $v_0 = 125$ ft/sec and $h = 5$ ft.

2. Plot the trajectory of the ball until the ball hits level ground, using a parametric plot. Start the parameter t at 0 and solve the equation $y = 0$ to find the time the ball hits the ground. Find the horizontal distance traveled by the ball.

3. What is the shape of the graph? Eliminate t, and determine the trajectory by giving y as a function of x.

   HINT: Solve the equation `x=x1` for t and substitute for t in `y1` using the `subs` command.

   NOTE: Now execute `restart;` and re-input the expressions `x1` and `y1` into Maple but with the initial angle θ, the initial speed v_0 and the time t as free variables. Keep $g = 32$ ft/sec^2 and $h = 5$ ft as constants.

4. Suppose the home run fence is 12 feet high and 350 feet from home plate. What is the minimum velocity at which the ball must leave the bat so that the ball barely clears the home run fence?

   HINT: *Be careful:* Do not assume any particular value of the angle θ. In fact your strategy should be as follows. First, eliminate t as you did in part 3 (but keeping v_0 and θ as free variables). Then substitute $x = 350$ and $y = 12$ into the resulting equation and solve for the speed v_0 in terms of the angle θ. (You may need a plot to determine which solution for v_0 is positive.) Finally, minimize v_0 as a function of θ.

5. Now assume air resistance acts on the ball. Air resistance acts in the opposite direction to the velocity of the ball and its magnitude is proportional to the speed. Newton's Laws lead to the equations

   $$\frac{d^2x}{dt^2} = -k\frac{dx}{dt} \quad \text{and} \quad \frac{d^2y}{dt^2} = -g - k\frac{dy}{dt}$$

   Here, k is a positive friction constant, which will be given later. Execute `restart` and solve these equations for x and y (using `dsolve` with initial conditions).

6. To check for consistency, take the limit of your solution as $k \to 0$ and see if your result agrees with the solution for x and y without air resistance.

7. Repeat parts 2 and 4, taking into account air resistance with $k = 0.1$. Compare your plots and your answers to your results without air resistance.

   NOTE: In part 4, the formula for v_0 as a function of θ is ugly (involving Lambert W functions) and its derivative is uglier. You would not want to find these by hand.

Related Activities: See Stewart's Applied Projects: *Which is Faster, Going Up or Coming Down?* and *Calculus and Baseball*.

12.26 Parachuting

BACKGROUND: Review systems of differential equations with initial conditions from Section 8.4. Recall the motion of an object falling with air resistance can be found from Newton's law, *Mass × Acceleration = Force* where the force is the sum of the downward force of gravity and the upward "drag" force of the air resistance. In symbols, this equation is

$$m\frac{d^2y}{dt^2} = -mg - k\frac{dy}{dt}$$

Here, m is the mass, y is the altitude (up is positive), $g = 9.8\frac{m}{sec^2}$ is the acceleration of gravity and k is the is the (positive) drag coefficient. Notice that the velocity $v = \frac{dy}{dt}$ is negative (since the object is falling). So the drag term, $-kv$, is positive (or up) thus tending to slow down the fall. You should separate the second order differential equation into a system of two first order equations:

$$\frac{dy}{dt} = v \quad \text{and} \quad m\frac{dv}{dt} = -mg - kv$$

ASSIGNMENT: A sky-diver, weighing 75 kg., jumps from a plane at an altitude of 3000 meters and free falls for T_1 seconds before pulling the rip chord. The parachute takes 4 seconds to open and then the sky-diver falls with a fully open parachute. Find the maximum time, T_1, the sky-diver can free fall before pulling the rip chord and still have a "gentle" landing. While the parachute is opening assume the drag coefficient is $k = 25$kg/sec. When the parachute is fully opened, assume $k = 110$kg/sec.
A landing is defined to be "gentle" if the velocity on impact is less than the impact velocity of an object dropped (free-fall) from a height of 4 meters.

0. Find the maximum velocity `Vgentle` for a gentle landing.

1. Find the altitude `Y1` and velocity `V1` at time `T1` after the first stage of free-fall. These are the initial conditions for the second stage.

2. Find the altitude `Y2` and velocity `V2` at time `T2` = `T1+4` after the second stage when the parachute is opening. These are the initial conditions for the third stage.

3. Find the time `T3` when the sky diver hits the ground and the velocity `V3` at this time. `V3` should be a function of `T1` which you can then equate to the gentle landing velocity and solve for the maximal safe `T1`. Plot the altitude and velocity as functions of time.

4. What is the landing velocity if the sky diver freefalls for a second less or more than the maximal safe time. Comments? It may be helpful to plot `V3` as a function of `T1` to see if there is a terminal velocity.

12.27 Radioactive Waste at a Nuclear Power Plant

BACKGROUND: Review first order, linear differential equations.

ASSIGNMENT: A nuclear power plant produces a waste product that is a radioactive isotope, called A. The isotope A has a half-life of 8 years and splits into radioactive isotopes B and C. 45% of the weight of A becomes B and 55% becomes C. Thus, for example, if 100 kilograms of A decay there will be 45 kilograms of B and 55 kilograms of C. Isotope B has a half-life of 15 years while isotope C has a half-life of 20 years and they decay into nonradioactive by-products. Answer the following questions.

1. Find the three decay constants.

2. Suppose you start with 300 kilograms of isotope A. What is the maximum amount of isotope B that will be present, and when will this occur? Answer the same question for isotope C.

3. When the power plant was first turned on, there was no isotope A, B, or C present. If the power plant operates so that it produces isotope A at the constant rate of 30 kilograms per year, what are the largest amounts of isotopes A, B and C that will be present, and when will these occur?

4. Federal safety requirements say that the reactor can never have on hand more than 600 kilograms of isotope A, 500 kilograms of isotope B, or 800 kilograms of isotope C. What is the maximum rate at which the power plant can produce isotope A without violating the federal regulations?

HINT 1: If a radioactive isotope is being produced by some source at the same time as it is decaying, how does that alter the differential equation for the rate of change of the amount of this isotope?

HINT 2: Be sure to plot A, B, and C, in order to ascertain if they have the qualitative behavior you expect.

HINT 3: In problem 4, repeat the computations of problem 3, but assume the power plant produces isotope A at the constant rate of R kilograms per year.

12.28 Visualizing Euler's Method

BACKGROUND: Review tangent lines and differential equations. Euler's method, constructs a numerical solution $y = F(x)$ on an interval $a \leq x \leq b$ of a differential equations of the form:
$$\frac{dy}{dx} = f(x,y) \quad \text{with the initial condition} \quad y(a) = y_0$$
NOTE: If $f(x,y)$ is independent of y, then the differential equation becomes $y' = f(x)$ and the solution is just the antiderivative $y = y_0 + \int_a^x f(x)\,dx$. So Euler's method provides a numerical antiderivative.

Euler's method starts by dividing the interval into n subintervals (called steps) each of width $h = \frac{b-a}{n}$ (called the step size) and naming the points $x_i = a+ih$, for $i = 0, 1, 2, ..., n$. Thus $x_0 = a$ and $x_n = b$. At each point $x = x_i$, the tangent line to $y = F(x)$ is $y = F(x_i) + F'(x_i)(x - x_i)$. However, from the differential equation, $F'(x_i) = \frac{dy}{dx} = f(x_i, F(x_i))$. So the tangent line becomes $y = F(x_i) + f(x_i, F(x_i))(x - x_i)$. Starting with the initial condition $F(x_0) = y_0$, Euler's method recursively approximates $F(x_{i+1})$ by
$$y_{i+1} = y_i + f(x_i, y_i)\,h \quad \text{for} \quad i = 0, 1, 2, ..., n-1$$
which is the linear approximation except that $F(x_i)$ has been replaced by y_i, found in the previous step. Euler's method can be implimented in Maple by using a do loop as shown in the following example. Be sure to try it!

EXAMPLE: Use Euler's method to approximate the solution to $\frac{dy}{dx} = y - 3x$ with $y(0) = 2$ on the interval $[0, 4]$ using 8 intervals. Plot the direction field, the exact solution and Euler's approximation in the same plot.

SOLUTION: Define the function, the interval, the number of steps and the initial condition. Then compute the step size.

```
> f:=(x,y)->y-3*x;
> a,b:=0,4; n:=8;
> x[0], y[0]:=0, 2;
> h:=(b-a)/n;
```

Now apply Euler's method:

```
> for i from 0 to (n-1) do
> x[i+1]:=x[i]+h;  y[i+1]:=evalf(y[i]+f(x[i],y[i])*h);
> end do;
```

The Euler approximation is plotted using a point plot:

```
> plotlist:=[seq([x[i],y[i]],i=0..n)];
> p1:=plot(plotlist, 0..4, -10..10, thickness=3): p1;
```

The direction field and the exact solution are plotted with the DEplot command in the DEtools package and the graphs are combined using the display command in the plots package. We need the packages, the differential equation and the initial condition.

```
>   with(plots):   with(DEtools):
>   deq:=diff(Y(X),X)=f(X,Y(X));
>   init:=[[0,2]];
```
NOTE: We have to use capital X and Y since x and y are already used for the Euler approximation.
```
>   p2:=DEplot(deq,Y(X), X=0..4, Y=-10..10, init):   p2;
>   display(p1,p2);
```
In this plot, notice that each straight line segment in the Euler approximation has a slope which agrees with the direction field at the *left* endpoint of the line segment. Also notice that the Euler approximation is above (an over estimate for) the true solution. You will be asked to explain these facts in the exercises.

ASSIGNMENT:

1. Use Euler's method to approximate the solution to $\frac{dy}{dx} = \frac{x}{y+4}$ with $y(-2) = 1$ on the interval $[-2, 4]$ using 3 intervals.

2. Find the exact solution using `dsolve`. (Since the equation is separable, you should be able to check this by hand.) Plot the direction field, the exact solution and Euler's approximation in the same plot.

3. Explain why each straight line segment in the Euler approximation has a slope which agrees with the direction field at the *left* endpoint of the line segment.

4. Notice that here the Euler approximation is below (an under estimate for) the true solution, while in the earlier example the Euler approximation was an over estimate. Explain this fact based on whether the solution is increasing or decreasing or concave up or concave down.

5. Repeat Exercises 1 and 2 but with $n = 12$. What happened to the graph of the Euler approximation?

6. Euler's method is usually too inaccurate to be of practical use. A significant improvement is the following, called the modified Euler's method (or Huen's method, or second order Runge-Kutta)

$$y_{i+1} = y_i + \frac{k_1 + k_2}{2} h \quad \text{where } k_1 = f(x_i, y_i) \text{ and } k_2 = f(x_i + h, y_i + hk_1)$$

which averages the slopes at x_i and x_{i+1}. Write a `do` loop to implement this method. Apply it to the Example and to Exercises 1 and 2 using $n = 3$ and $n = 12$. Plot the exact solution together with the Euler approximation and the Runge-Kutta approximation.

12.29 The Brightest Phase of Venus

BACKGROUND: Study trigonometry, polar and spherical coordinates and max/min problems and area. (This is the hardest project.)

ASSIGNMENT: The brightness of Venus is proportional to the area of the visible illuminated portion of Venus and inversely proportional to the square of the distance from the Earth to Venus. From the figure, note that, as the angle α increases from 0 to π, the area of the visible illuminated portion of Venus increases, which tends to increase the brightness of Venus. However, the distance d from the Earth to Venus also increases, which tends to decrease the brightness of Venus. For some angle α, between 0 and π, Venus will appear brightest. Find this brightest phase of Venus (i.e., the angle α).

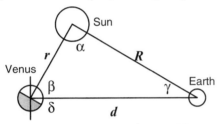

Here, $r = 67$ is the distance from the Sun to Venus and $R = 93$ is the distance from the Sun to the Earth, where the unit of distance is a million miles. As mentioned above, the brightness is

$$B = k \frac{\text{Area of the visible illuminated portion of Venus}}{d^2}$$

where k is a proportionality constant. We want to express B in terms of α. Here are some suggestions that may help.

1. The distance d can be found from the law of cosines:

$$d^2 = r^2 + R^2 - 2rR\cos(\alpha)$$

2. Choose a coordinate system with Venus at the origin, the Earth on the positive x-axis and the Sun in the first quadrant in the xy-plane. Then the z-axis points out of the paper. In terms of the polar (or spherical) coordinate θ, the sun illuminates from $\theta = -\delta$ to $\theta = \pi - \delta$. (See the figure.) However, the earth can only see the portion with $-\delta \leq \theta \leq \pi/2$. Venus is actually a sphere (of radius a), but from the Earth we see the disk which is the projection of the sphere into the yz-plane. The area we actually want is the illuminated portion of this disk. In spherical coordinates, the surface of Venus is

$$\left.\begin{array}{l} x = a\sin\phi\cos\theta \\ y = a\sin\phi\sin\theta \\ z = a\cos\phi \end{array}\right\} \quad \text{for} \quad \left\{\begin{array}{l} 0 \leq \phi \leq \pi \\ -\pi \leq \theta \leq \pi \end{array}\right.$$

So the projection of the illuminated portion on the yz-plane is

$$\left. \begin{array}{l} y = a \sin\phi \sin\theta \\ z = a \cos\phi \end{array} \right\} \quad \text{for} \quad \left\{ \begin{array}{l} 0 \leq \phi \leq \pi \\ -\delta \leq \theta \leq \pi/2 \end{array} \right.$$

The edges of this illuminated projection are two parametric curves:
When $\theta = \pi/2$ the edge is the semicircle $y = a \sin\phi$, $z = a \cos\phi$.
When $\theta = -\delta$ the edge is the semi-ellipse $y = -a \sin\phi \sin\delta$, $z = a \cos\phi$.
Eliminate the parameter ϕ to find equations for the edges and integrate to find the area as a function of the angle δ.

3. It remains to express δ in terms of α. This can be done with the complementary angle relation $\beta + \delta = \dfrac{\pi}{2}$ and the law of sines $\dfrac{\sin\beta}{R} = \dfrac{\sin\alpha}{d}$.
NOTE: In solving for β, remember arcsin always returns an angle between $-\pi/2$ and $\pi/2$. So you need two different formulas for $\beta \leq \pi/2$ and $\beta > \pi/2$. How is $\beta > \pi/2$ expressed in terms of R, r and d? In terms of α? This leads to two formulas for δ.

4. Combine these results to express the brightness B in terms of the one variable α. You again need two formulas for B, one for $\beta > \pi/2$ and one for $\beta < \pi/2$. Use Maple to maximize B over the interval $0 \leq \alpha \leq \pi$. From your answer, which is more important, increasing the illuminated portion of Venus or decreasing the distance from the Earth to Venus?

Appendix A

The Maple Interfaces

This chapter will discuss which Maple interface to use with this book, how to configure your worksheets and additional resources in the Standard Interface and at Maplesoft's website.

A.1 Selecting an Interface and Mode

Maple comes with two interfaces: the Standard Interface (Maple 12 on the Start menu) and the Classic Interface (Classic Worksheet Maple 12 on the Start menu). Further the Standard Interface can be used in either Worksheet Mode or Document Mode.

Whenever you start a new worksheet you need to choose an *interface* and a *mode*. Pick it and stick with it. There are incompatibilties in going back and forth. The Standard Interface is being improved with each new edition of Maple and has many bells and whistles not available in the Classic Interface. The Classic Interface is a little faster and avoids some problems with copy and paste and with hidden characters that are present in the Standard Interface. However, the Classic Interface is not being maintained and does not have any of the new features available in the Standard Interface (which will be described later in this appendix). The Worksheet Mode has a linear structure in which you enter a Maple command on an input line and the result appears directly below it on an output line. The Document Mode is designed to produce beautiful (printable or online) documents in which you can position various elements (text, plots, tables, spreadsheets, buttons, menus, etc.) wherever you want on each page, with the Maple commands either visible or hidden.

For doing the assignments in this text, we recommend working with either the Classic Interface or the Worksheet Mode of the Standard Interface. Do not use the Document Mode; it is unnecessarily complex. For each of these interfaces, there are configuration options that we recommend. These are described below. These can be set once and for all so that you do not need to worry about them unless you go to a different computer.

A.2 The Classic Interface

In the Classic Interface, there are three options to be set for optimal performance. These are explained below. However the recommended settings are obtained by using the following sequences of menu selections:

- FILE > PREFERENCES > GENERAL > AUTOSAVE > ENABLE AUTOSAVE EVERY 3 MINUTES
- FILE > PREFERENCES > I/O DISPLAY > ASSUMED VARIABLES > NO ANNOTATION
- FILE > PREFERENCES > I/O DISPLAY > INPUT DISPLAY > MAPLE NOTATION

After making these three selections, click on Apply Globally to make the changes permanent.

AutoSave: When you are doing a long complicated computation, Maple will sometimes lock up. If you have not saved recently, you may lose a lot of work. The AutoSave feature will save most of your work. Every 3 minutes is a reasonable compromise between saving too frequently and not frequently enough. If your worksheet is named "myfile.mws", the autosaved file will be named "myfile_MAS.mws".

Assumed Variables: Maple has the ability to make assumptions about quantities and then use them. For example, if you execute

> assume(a>b, b>0);

then Maple can answer such questions as

> is(sqrt(a^2-b^2):: real);

$$true$$

However, if the ASSUMED VARIABLES option is set to TRAILING TILDES, then any formula involving variables with assumptions will have tildes following the those variables, to remind you there are assumptions. For example

> b^2-4*a*c;

$$b^{\sim 2} - 4\, a^{\sim}\, c$$

When the formula gets long, the tildes make the formula longer and can make it hard to read. However, if the ASSUMED VARIABLES option is set to NO ANNOTATION, then the tildes are eliminated. For example

> b^2-4*a*c;

$$b^2 - 4\, a\, c$$

In this text we will make no annotation on the assumed varibles.

Input Display: Maple input can be entered either in 1-dimensional notation such as

> x/(x^2+1);

or in 2-dimensional notation such as

> $\dfrac{x}{x^2+1}$

The 1-dimensional notation is called MAPLE NOTATION. The 2-dimensional notation is called STANDARD MATH NOTATION. You type exactly the same thing (x/(x^2+1)) to get both of these, however, the input will appear as specified by the INPUT DISPLAY option. To change this setting temporarily for one line, when the cursor is at a new prompt, select INSERT > STANDARD MATH INPUT or INSERT > MAPLE INPUT as appropriate. If you have already typed an input line and want to convert it to the other format, highlight the input with the mouse and select FORMAT > CONVERT TO > STANDARD MATH INPUT or FORMAT > CONVERT TO > MAPLE INPUT as appropriate.

The advantage of the 2-dimensional input is that it looks like standard math. The advantage of the 1-dimensional input is that it looks like what you actually typed. Further it is easier to type corrections to the Maple input than to the Standard input. (Try it!) In this book, we have typed everything in Maple notation, because it is then easier for you to know what to type. Further, we recommend that you also use Maple notation, because it is then easier for your instructor to know what you typed when grading your worksheets. However, most instructors will also accept Standard notation if you really prefer it.

Input Palettes: As an aid to figuring out what to type to get a particular input, Maple provides a set of palettes which may be used to select an input format at the click of a mouse. These can all be displayed by selecting VIEW > PALETTES > SHOW ALL PALETTES. You can click any button on the Symbol, Expression, Matrix or Vector palettes and the symbol will be inserted into the worksheet. If, for example, you click the a^b button on the Expression palette, then in Maple input you will see

> (%?^%?);

while in Standard input you will see

> ?$^?$

Each %? or ? denotes a place where you can type any quantity you wish. You can move between input regions using ⟨TAB⟩ or ⟨SHIFT-TAB⟩. As you move through this book you may find some of these palettes useful. Use them as you wish. Experienced Maple users tend to remember the syntax.

A.3 The Standard Interface

In the Standard Interface, there are four options to be set for optimal performance. These are explained below. However the recommended settings are obtained by using the following sequences of menu selections:

- TOOLS > OPTIONS > GENERAL > AUTOSAVE > AUTOSAVE EVERY 3 MINUTES
- TOOLS > OPTIONS > DISPLAY > ASSUMED VARIABLES > NO ANNOTATION

- TOOLS > OPTIONS > INTERFACE > DEFAULT FORMAT FOR NEW WORKSHEETS > WORKSHEET

- TOOLS > OPTIONS > DISPLAY > INPUT DISPLAY > MAPLE NOTATION

After making these four selections, click on Apply Globally to make the changes permanent.

AutoSave: See the discussion above under The Classic Interface, except that if your worksheet is named "myfile.mw", the autosaved file will be named "myfile_MAS.bak" and you need to change the extension to ".mw" before you can open it.

Assumed Variables: See the discussion above under The Classic Interface.

Default Format for New Worksheets: When you open a new worksheet using FILE > NEW, you get a choice between the DOCUMENT MODE and the WORKSHEET MODE. Always choose WORKSHEET MODE. The Document allows for fancy layout on the page (which is unnecessary for this course) and is significantly more complicated to use.

When you open a new worksheet by starting Maple, by clicking on the open icon on the toolbar or by typing ⟨CTRL-N⟩, the new worksheet is opened using the mode specified by the DEFAULT FORMAT FOR NEW WORKSHEETS option. So set this to WORKSHEET.

Input Display: Maple input can be entered either in 1-dimensional notation such as

> x/(x^2+1);

or in 2-dimensional notation such as

$$> \frac{x}{x^2+1}$$

The 1-dimensional notation is called MAPLE NOTATION. The 2-dimensional notation is called 2-D MATH NOTATION. You type almost the same thing to get both of these, x/(x^2+1) in Maple notation and x/x^2 right-arrow +1 in 2-D notation. Here "right-arrow" means the cursor key.

Which input will appear is specified by the INPUT DISPLAY option. To change this setting temporarily for one line, when the cursor is at a new prompt, select INSERT > 2-D MATH or INSERT > MAPLE INPUT as appropriate. If you have already typed an input line and want to convert it to the other mode, click on the input with the mouse and select FORMAT > CONVERT TO > 2-D MATH INPUT or FORMAT > CONVERT TO > 1-D MATH INPUT as appropriate.

The advantage of the 2-dimensional input is that it looks like standard math. The advantage of the 1-dimensional notation is that it looks like what you actually typed. Further it is easier to type corrections to the Maple input than to the Standard input – there are many more hidden characters. (Try it!) In this book, we have typed everything in Maple notation, because it is then easier for you to know what to type. Further, we recommend that you also use Maple notation, because it is then easier for your instructor to know what

A.3. THE STANDARD INTERFACE

you typed when grading your worksheets. However, most instructors will also accept Standard notation if you really prefer it.

Input Palettes: As an aid to figuring out what to type to get a particular input, Maple provides a set of palettes which may be used to select an input format at the click of a mouse. The first time you open Maple, there is a list of palettes along the left side as shown below.

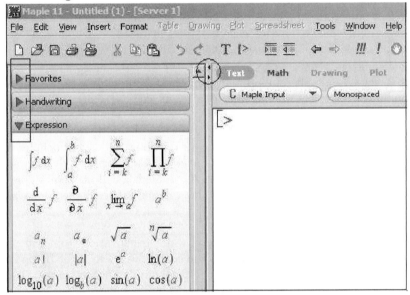

You can hide or reveal the list of palettes by clicking on the small left and right arrows shown in a circle in the above figure. You can expand or collapse each palette by clicking on the large right or down arrows shown in a rectangle in the above figure. You can click any button in any palette and the symbol will be inserted into the worksheet. If, for example, you click the a^b button on the Expression palette, then in Maple input you will see

>   (a)^(b);

while in Standard input you will see

> a^b

Each a or b denotes a place where you can type any quantity you wish. You can move between input regions using ⟨TAB⟩ or ⟨SHIFT-TAB⟩. As you move through this book you may find some of these palettes useful. Use them as you wish. Experienced Maple users tend to remember the syntax.

Tutors and Assistants: The Standard interface provides easy access to a collection of Tutors and Assistants. Just click on TOOLS > TUTORS or TOOLS > ASSISTANTS and select the one you want. Each of these is a pop-up window (called a maplet) which helps you through the computations on some topic. As you work through this book, periodically look at the lists of Tutors and Assistants to see if there is one on the topic you are learning.

A.4 The Maplesoft Website

There are many additional resources available at the Maplesoft website:
http://maplesoft.com/
Some are free and some are for sale.

The primary free resource is the Maple Application Center at:
http://maplesoft.com/applications/
It offers numerous Maple worksheets, packages and maplets in such diverse areas as mathematics, physics, chemistry and engineering including the `VecCalc` package by Belmonte and Yasskin for multivariable vector calculus.

The commercial resources are available at the Maple Connect site:
http://www.maplesoft.com/products/thirdparty/
including the Maplets for Calculus. Maplesoft also has a site dedicated to students:
http://maplesoft.com/students/
On this page, we point out the Maple 12 Calculus Kit which includes a copy of Maple 12, the Precalculus Study Guide, the Calculus Study Guide and the Maplets for Calculus. The Calculus Study Guide is a collection of Maple worksheets on various topics in single variable calculus. The Maplets for Calculus, which are described in detail at:
http://m4c.math.tamu.edu/
is a collection of maplets (Maple applets) which teach various topics in calculus whereas the built-in Tutors simply do the computations for you. If you already own a copy of Maple 12 (e.g., it was packaged with your copy of your Calculus textbook), then the other parts are available individually.

Index

abs, 2, 196
allvalues, xii
and, 149
angle of inclination, 65
animate, 203
ant, 201
ApproximateInt, 94
arc length, 109, 114, 121, 124, 218
arccos, 2
arccot, 2
arccsc, 2
arcsec, 2
arcsin, 2
arctan, 2, 11, 65, 108
area, 71, 93, 109, 120, 121, 210, 217, 233
    p-ball, 212
    circle, 3
    polar, 124
    to tangent line, 217
args, 149, 158
assignment, xi, 3, 125
Assistants, 239
Assumed Variables, 236, 238
asymptote, 76, 85
@, 17
AutoSave, 236, 238
average value, 117

baseball, 227
Birthday Problem, 154
blimp, 208
bouncing ball, 226
break, 149, 153

center of mass, 213, 214
chain rule, 61, 210

change of variables, 98
changevar, xiii, 99, 105, 108
circle
    data points, 51
    parametric, 42
Classic Interface, 235, 236
colon (:), xi
composition, 17, 210
concavity, 217
conditional, 149
constant of integration
    in differential equation, xiv, 125
    in indefinite integral, xiii, 98
constrained, 8, 13, 59
convert
    ln, 99
    parfrac, 101, 105
    polynom, xiv, 143, 145
Copy, 177
cos, 2
cot, 2
critical point, 80, 86
csc, 2
⟨Ctrl-Delete⟩, xiv
curve fitting, 37, 41, 51
    approximate, 40
    exact, 39
cycloid, 218

D, xiii, 56, 65, 126, 158, 201, 210
data set, 37, 122
    random, 52
decreasing, 87
Default Format, 238
delete, xiv
denom, 77
DEplot, 125, 127, 129, 133

derivative, 54, 56, 210
    higher, 57, 59, 144
    implicit, 61, 64
    of expression, 56, 58, 64
    of function, 56, 63
`DEtools` package, 127, 129, 130, 133
`Diff`, xiii, 58
`diff`, xiii, 58, 61, 64, 65, 75, 126, 210
differential equation, 125, 133, 230
    direction field, 127
    Euler's method, 128, 231
    first order, 125, 127
    linear, 125
    nonlinear, 128
    numerical solutions, 128
    second order, 126
    separable, 226
    system, 130, 133, 229
`Digits`, 11, 199
dipstick, 211
direction field, 125, 127, 130
`discont`, 7
`display`, xiii, 22, 35, 39, 51, 88
`do`, 149, 152
Document Mode, 235
donut, 123, 216
`dsolve`, xiv, 125, 133, 226
    initial condition, 126
    `numeric`, 125, 129, 131, 133, 134

e, xi
`elif`, 151
`else`, 149, 151
endpoint, 86
⟨ENTER⟩, xi, 1, 5
epicycloid, 218
equation, xi, 14, 25, 31, 125, 201
equations
    system of, 30, 79
error estimate
    integral test, 140
Euler's method, 128, 231
`eval`, xii, 15, 31, 32, 56, 58, 64, 99, 108, 173
`evalf`, xi, 2, 3, 10, 27, 31, 77, 106
    `Int`, 98, 120, 124

`RootOf`, 188
execute command, xi, 1
EXECUTE WORKSHEET, xiv, 4, 9
`exp`, xi, 2, 76
`expand`, xi, 4, 10, 12, 96, 199
expression, xi, 4, 14, 31, 54, 75, 125, 200
    derivative, 56, 58
    integral, 97
extremum, 79
    absolute, 80, 86
    local, 75, 84, 85

`factor`, xii, 10, 17, 33, 67, 199
`false`, 149
`fit[leastsquare]`, 41
focal length, 74
Folium of Descartes, 59
`for`, 149, 152
Fourier cosine expansion, 118, 124
Fourier series, 109, 117, 121
`fsolve`, xii, 25, 27, 31, 34, 60, 63, 69, 71, 75, 86, 120, 124, 187, 188, 199, 201, 202
    `complex`, 27
    with range, 28
function, xi, 14, 31, 54, 125, 198
    as procedure, 155
    derivative, 56, 201
    implicit, 59
    inverse, 44, 211

geometric series, 138, 226
`global`, 156
graphing, 73, 76, 84, 85
`grid`, 29

half-life, 230
happy face, 202
harmonic series, 138
help, xi, 5, 163
higher derivative, 144
    of expression, 59, 64
    of function, 57, 64
horizontal asymptote, 76
hyperbola, 197

parametric, 53
hypocycloid, 218

`if`, 149, 151
`ifactor`, 12
implicit derivative, 61, 64
implicit function, 59
`implicitplot`, xiii, 29, 36, 59, 62, 64, 67
`implicitplot3d`, 31
increasing, 87
inflection point, 76, 84, 85
initial condition, xiv, 126
Input Display, 236–239
`Int`, xiii, 93, 97, 105, 106, 120, 123, 124, 135, 145, 210
integral curve, 131
integral test, 138
    error estimate, 140
integration, 93, 95, 210
    applications, 109
    change of variables, 98, 105
    numerical, 102, 231
    partial fractions, 101, 105
    parts, 100, 105
    substitution, 98, 105
integration by parts, 100
intercept, 84, 85
interest, 224, 226
`intparts`, xiii, 100, 105, 108
inverse function, 37, 44, 211

JELL-O mold, 122

least squares fit, 41
left Riemann sum, 102
`leftbox`, 93, 105, 106
`leftsum`, 102, 105, 106, 122
letters, 206
`lhs`, xi, 202
`Limit`, xiii, 54, 55, 77, 97, 105, 135, 145, 198, 200
    of sequence, 137
limit comparison test, 138
linear approximation, 64
Lissajous figure, 43, 114, 123

list, 136, 201
`ln`, 2
`local`, 156
logic, 149
loops, 149, 152
Lotka-Volterra, 130, 134

max/min problems, 73, 81, 216, 233
maximum, 79, 87
    absolute, 80, 86
    local, 75, 84, 85
meteorite, 197
`middlebox`, 106
`middlesum`, 102, 105, 106
midpoint Riemann sum, 102
midpoint rule, 93
minimum, 79, 87, 217
    absolute, 80, 86
    local, 75, 84, 85
`module`, 149, 163
movie, 89

`nargs`, 149, 158
Newton's method, 69
`next`, 149, 153
`nops`, 158
normal line, 205
`not`, 149
nuclear power, 230
numerical integration, 102
`numpoints`, 35

`odeplot`, 129, 132, 133
optics, 74
`or`, 149
order of operation, 1

p-ball, 212
package
    `DEtools`, 127
    `plots`, 39
    `statplots`, 40
    `stats`, 40
    `student`, 105
    `Student[Calculus1]`, 94, 143
parabola, 214

data points, 122
    polar, 48
parachute, 200, 229
parametric curve, 37, 42, 206, 207, 218, 227
    inverse function, 45
parametric `plot`, xiii, 42, 218
`parfrac`, 101, 105
partial fractions, 101
partial sum, 140, 146
PASTE, 177
pension, 226
`%`, xi, 3
phase portrait, 125, 131
phase trajectory, 131
`Pi` (π), xi, 3
`piecewise`, 35
plot
    2D graph, 212
    2D implicit, 212
`plot`, xii, 6, 17, 20, 33, 34, 75, 135, 145, 210
    delete, 9
    equation, 14
    expression, 14
    function, 14
    inverse function, 44
    parametric, xiii, 42
    points, 37, 201
    `polar`, 47
    sequence, 136
    tangent line, 55, 62
`plot3d`, 23, 35
`plots` package, xiii, 22, 29, 39, 64, 93, 124, 129, 132
point of intersection, 66
polar curve, 37, 47, 124
`polarplot`, 47, 124
power relay, 196
`print`, 156
`proc`, 149, 156
procedure, 149, 155, 188
product rule, 61
programming, 149
PROJECTION, 8, 13
projects, 195

pulley, 72

radioactive waste, 230
`rand`, 52
ratio test, 138
related rates, 73, 83
REMOVE OUTPUT, xiv, 9
`restart`, xiv, 4
⟨RETURN⟩, 1
`return`, 149, 159
`rhs`, xi, 62, 125, 202
Riemann sum, 93, 105, 106
right Riemann sum, 102
`rightbox`, 93, 105, 106
`rightsum`, 102, 105, 106
rolling circle, 218
root test, 138
`RootOf`, xii, 27, 67, 188
Runge-Kutta, 128, 232

SAVE, xiv, 4, 191
`scaling`, 8, 13, 59
`scatterplot`, 40
`sec`, 2
Second Derivative Test, 217
semicolon (;), xi, 1
`seq`, 135, 145
sequence, 135, 145
series, xiv, 135, 137
    approximating, 140
    convergence tests, 138
⟨SHIFT-ENTER⟩, 150
Shylock, 224
`simplify`, xii, 5, 10, 56, 64, 108
`simpson`, 102, 105
Simpson's rule, 93, 102, 108
`sin`, 2, 9, 12
sky-diver, 229
slope, 13, 54, 62, 63, 65, 218
    matching, 88
Snell's Law, 86, 90
solution curve, 125, 130
`solve`, xii, 25, 31, 33, 34, 39, 64, 75, 187
`sqrt`, 2, 11
Standard Interface, 235, 237

statplots package, 40
stats package, 40
Stewart, ix
STOP sign, xi, 192
student package, xiii, 13, 93, 99, 100, 102, 105
Student[Calculus1], 94, 143
style=point, 6, 68
subs, xii, 15, 31, 32, 58, 63, 64, 77, 173, 199
substitution, 98
Sum, xiv, 93, 96, 105, 135, 137, 166
surface area, 109, 114, 121, 124, 216
system of equations, 30, 79

tan, 2, 11
tangent line, 57, 62–66, 69, 201, 205, 217, 231
taylor, xiv, 135, 143, 145
Taylor polynomial, 143, 147
    plot, 145
Taylor Remainder Formula, 144, 147
TaylorApproximation, 135, 143, 145
Texas, 51, 88, 107, 108, 213
Texas Aggie, 198
then, 149, 151
tin can, 82
torus, 216
trapezoid, 102, 105
trapezoid rule, 93, 102, 108, 213
trough, 71
true, 149
Tutors, 239

unapply, 19, 31, 33, 62
unassign, xi, 4

value, xiii, 55, 145
    changevar, 99, 108
    Diff, 58
    Int, 97, 105, 120, 123, 135, 210
    intparts, 100, 108
    leftsum, 122
    Limit, 77, 97, 105, 135, 137
    Sum, 96, 105, 135, 137, 166
variable, 14

velocity, 198, 200
Venus, 233
vertical asymptote, 76
view, 49
volume, 71, 109, 111, 216
    by cylinder, 113, 123
    by disk, 113
    by slicing, 111, 120, 122
    by washer, 113
    of revolution, 113, 120, 121, 211

waffle cone, 71
while, 149, 152
Worksheet Mode, 235